DNA-BASED MOLECULAR ELECTRONICS

Related Titles from AIP Conference Proceedings

To learn more about these titles, or the AIP Conference Proceedings Series, please visit the webpage **http://proceedings.aip.org**

DNA-BASED MOLECULAR ELECTRONICS

International Symposium on
DNA-Based Molecular Electronics

Jena, Germany   13 – 15 May 2004

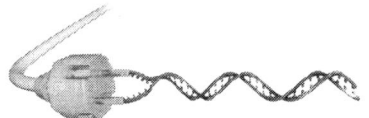

EDITOR
Wolfgang Fritzsche
IPHT, Jena, Germany

SPONSORING ORGANIZATIONS
DFG - Deutsche Forschungsgemeinschaft
IPHT - Institut für Physikalische Hochtechnologie
STIFT - Stiftung für Technologie, Innovation und Forschung Thüringen
FCI - Fonds der Chemischen Industrie

AMERICAN INSTITUTE OF PHYSICS

Melville, New York, 2004
AIP CONFERENCE PROCEEDINGS ■ VOLUME 725

Editor:

Wolfgang Fritzsche
IPHT-Institut für Physikalische Hochtechnologie
P. O. Box 100 239
D-07702 Jena
GERMANY

E-mail: fritzsche@ipht-jena.de

L.C. Catalog Card No. 2004111432
ISBN 0-7354-0206-X
ISSN 0094-243X
Printed in the United States of America

CONTENTS

APPENDICES

Preface

A future electronics based on molecular wires and devices represents an interesting but still distant option. The last years witnessed a great interest in DNA-based approaches in the molecular electronics field, and the meeting mirrored this development with contributions in areas like DNA modification, manipulation of DNA in electrical fields, or DNA superstructure nanotechnology. The use of DNA as "glue" to connect molecular units in order to establish molecular complexes, the metallization of DNA and the interesting world of DNA superstructures based on reciprocal exchange between DNA double helices or on Guanine quartets emerged as focus points. The results represent first real steps in the visionary direction of molecular electronics, and will contribute to further progress by converging experimental disciplines like physics, molecular biology, surface chemistry and microsystem technology.

Wolfgang Fritzsche

Scientific Committee of the Symposium

Cees Dekker (Delft)

Wolfgang Fritzsche (Jena)

Bernd Giese (Basel)

Ned Seeman (New York)

Itamar Willner (Jerusalem)

Masao Washizu (Tokyo)

CONTROL OF DNA POSITIONING

Adsorption of DNA molecule and DNA Patterning on Si substrate

Shin-ichi Tanaka, Masateru Taniguchi and Tomoji Kawai

The Institute of Scientific and Industrial Research, Osaka University, 8-1, Mihogaoka, Ibaraki, Osaka 567-0047, Japan, CREST JST

Abstract. DNA molecule is a candidate electrical material for molecular devices. However, in order to realize a DNA molecular device, it is necessary to combine characteristics of DNA with semiconductor technology. DNA molecule is adsorbed not on the SiH surface but on the SiO_2 surface by adding $MgCl_2$ to DNA solution. In addition, DNA molecule can be selectively adsorbed to SiO_2 surface in SiO_2/SiH pattern, which is fabricated using photolithography, and DNA patterning is made on Si substrate. Since DNA molecule can be adsorbed to Si substrate through Mg^{2+}, the adsorption of DNA molecule in SiO_2/SiH pattern is depended on the concentration of $MgCl_2$ and the difference of chemical property between SiO_2 surface and SiH surface. The optimum concentration of $MgCl_2$ in which DNA is selectively adsorbed to SiO_2 surface was 0.1 mM.

1. INTRODUCTION

Semiconducting devices and magnetic memories are the fundamental paradigm of electronics at the present time. However, their miniaturization and acceleration will reach their physical limit in the future. In order to develop a further ultrahigh performance device, it is necessary to develop a system that can work at nano meter size.

In the construction of molecular device, there are two key technologies. One is the "Top down" technology which fabricates a material in submicrometer or nanometer size, and the other is "Bottom up" technology, which means that a wide range of molecular construction is formed by self-assembly and molecular recognition and that a functional supermolecule and/or macromolecule is synthesized by using organic chemical method.

Silicon is the most important material in the field of electronics at present time and numerous studies have ever been conducted. Therefore, since micro-processing technology of silicon substrate such as photolithography and electron beam lithography has been established, it is possible to fabricate a pattern in nanometer or micrometer size [1, 2]. On the other hand, the DNA molecule is expected as an electrical material for molecular devices, since each base molecule in DNA possess an inherent energy level (the highest occupied molecular orbital (HOMO) and the lowest unoccupied molecular orbital (LUMO)) and the electric state of DNA can be controlled by base sequence. In addition, since there are many reports that DNA molecule is a wide gap semiconductor [3-20], the conductivity of DNA can be

CP725, *DNA-Based Molecular Electronics: International Symposium*, edited by W. Fritzsche
© 2004 American Institute of Physics 0-7354-0206-X/04/$22.00

controlled by doping with some chemical species such as I_2 and $FeCl_3$ like conductive polymers [21]. Moreover, DNA can form self-assemble structure on the atomically flat substrate [22-24]. However, in order to realize a DNA molecular device, it is necessary to combine the characteristics of DNA with semiconductor technology.

2.   ADSORPTION OF DNA MOLECULE ON SI SUBSTRATE

DNA could not be adsorbed to Si substrate, when the DNA solution without $MgCl_2$ was used. Then, we referred the literature [25] that DNA (pUC18-DNA) was attached on Si substrate by added $MgCl_2$ to DNA solution.

Si substrate was flattened by RCA method [26]. A unique triangle pattern on the RCA treated Si (111) surfaces is observed in atomic force microscope (AFM) images, as shown in Fig. 1 (a) and (b). The height of the step in both surfaces was 0.5 nm. The average roughnesses of the SiO_2 (111) surface and SiH (111) surface were 0.3 nm and 0.1 nm, respectively, and these surfaces were sufficiently flat to observe DNA.

The Poly(dG)·Poly(dC) (250 µg/ml, Amersham Pharmacia Biotech.) solution containing $MgCl_2$ (0.5 mM, Sigma-Aldrich Japan K.K.) was dropped onto the SiO_2 and SiH surface and then blown off after 1min. The DNA molecule was attached and formed a network structure on the SiO_2 (111) surface (Fig. 1 (c)). Conversely, DNA did not attach to the SiH (111) surface and the roughness of this surface was changed from 0.5 to 1 nm (Fig. 1(d)).

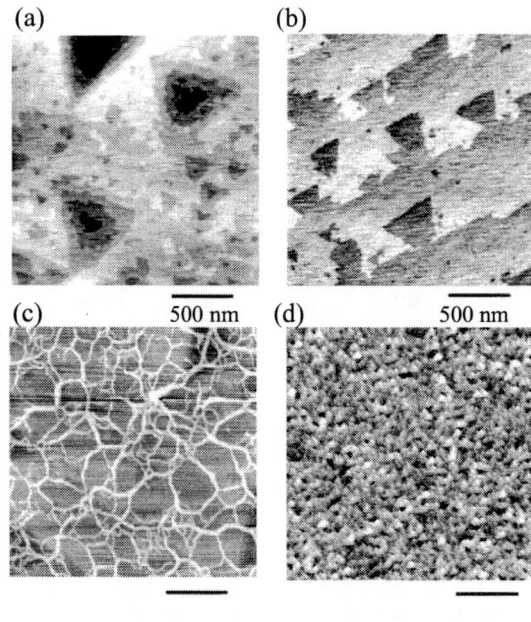

FIGURE 1. AFM images of (a) SiO_2 (111) surface and (b) SiH (111) surface. The DNA network structure by spreading Poly(dG-dC)·Poly(dG-dC) (250 µg/ml) solution with $MgCl_2$ (0.5 mM) on (c) SiO_2 (111) surface and (d) SiH (111) surface. (2 µm×2 µm)

In the solution, Mg^{2+} attaches to the negative charge on the DNA backbone, and the DNA is charged positively. Since coordination bond can be formed between and Mg^{2+} and the unshared electron pair of oxygen atoms on the SiO_2 surface, Mg^{2+}-DNA can be adsorbed to the SiO_2 surface [27-30]. However, the DNA cannot attach to the SiH (111) surface, because the hydrogen atoms and Mg^{2+} ion do not form coordination bonds. The DNA microwiring pattern can be controlled by fabricating the SiO_2/SiH micro pattern.

3   DNA PATTERNING ON SI SUBSTRATE

In fabrication of SiO_2/SiH pattern, Si substrate treated with RCA method was annealed in the presence of oxygen in order to grow the SiO_2 layer. The thickness of the SiO_2 layer was about 50 nm. The fabrication of SiO_2/SiH pattern was performed by using an image reversal process and reactive ion etching (RIE), as shown in Fig. 2.

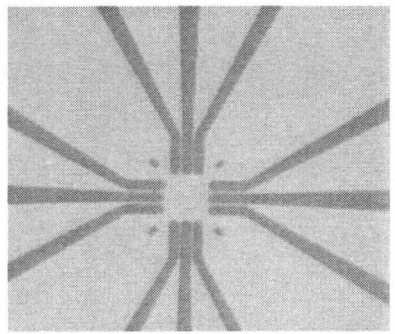

50 μm

FIGURE 2. The optical microscope image of fabricated SiO_2/SiH pattern.

We examined $MgCl_2$ concentration dependence of the adsorption of the DNA molecule on the SiO_2/SiH pattern in the 0.1 to 5.0 mM region. These results are shown in Fig. 3. When the $MgCl_2$ concentration was 0.1 mM, a string-like molecule whose the height was 0.8-1.2 nm and the length was 300-500 nm, was observed on the SiO_2 surface (Fig. 3 (a)). Since the height of the observed molecule was approximately equal to the reported average height of a single DNA (1 nm) [31], this is probably a single DNA. On the other hand, DNA was not observed on the SiH surface (Fig. 3 (b)). When the concentration of $MgCl_2$ was increased above 1.0 mM, DNA formed a characteristic network structure on the SiO_2 surface whose height of the DNA network structure was 0.8-1.2 nm (Fig. 3 (c)). On the other hand, a DNA agglomerate was observed on the SiH surface whose height was 0.8-1.0 nm (Fig. 3 (d)). In the solution, the electrostatic bond is formed between Mg^{2+} and DNA, and DNA can be adsorbed to SiO_2 surface. As the number of Mg^{2+} on DNA increases with increasing $MgCl_2$ concentration, more DNA can be adsorbed to the SiO_2 surface. However, since Mg^{2+}-DNA is liable to aggregate each other in the solution, the aggregation of DNA was facilitated by increasing MgCl$_2$ concentration. Therefore, the aggregated DNA was observed on the SiH surface in high MgCl$_2$ concentration, though Mg^{2+}-DNA is unable to be absorbed on the SiH surface. From these results, the number of DNA adsorbed on the SiO$_2$ and SiH surface were dependent on the concentration of MgCl$_2$, and DNA molecules can be selectively adsorbed to the SiO$_2$ surface by preparing the solution with an MgCl$_2$ concentration of 0.1 mM.

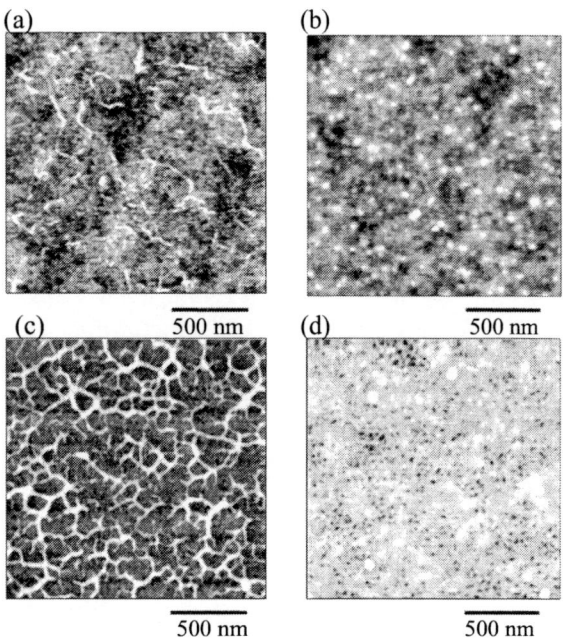

FIGURE 3. AFM images of the (a) SiO$_2$ (111) surface (0.1 mM MgCl$_2$), (b) SiH (111) surface (0.1 mM MgCl$_2$), (c) SiO$_2$ (111) surface (5.0 mM MgCl$_2$) and (d) SiH(111) surface (5.0 mM MgCl$_2$). The adsorption of Poly(dG-dC)·Poly(dG-dC) on SiO$_2$/SiH pattern was depended on MgCl$_2$ concentration. (2 μm_2 μm)

We examined the time dependence of the adsorption of a DNA molecule in SiO$_2$/SiH pattern by varying time intervals (1, 5, 10 min.) between the dropping and blowing off DNA solution containing 0.1 mM MgCl$_2$. When the time interval was varied, the number of DNA adsorbing on the SiO$_2$ did not changed. This indicates that the adsorption of the DNA molecule was influenced not by the difference of adsorption rate but by the difference of chemical property between the SiO$_2$ surface (hydrophobicity) and the SiH surface (hydrophilicity).

We performed fluorescence microscope observation in order to confirm whether the molecule in the SiO$_2$/SiH pattern was DNA. Then, DNA was stained with Yo-Pro. Fluorescence and optical images are shown in Fig. 4. In fluorescence microscope observation, we observed the fluorescing string-like molecule as shown in Fig. 4 (a). The string-like molecules observed in fluorescent images were similar to those in AFM images. On the other hand, no fluorescing molecule was observed on the SiH surface. Thus, we succeeded that DNA molecule was selectively adsorbed to the SiO₂ surface in the SiO₂/SiH pattern.

(a)   (b)

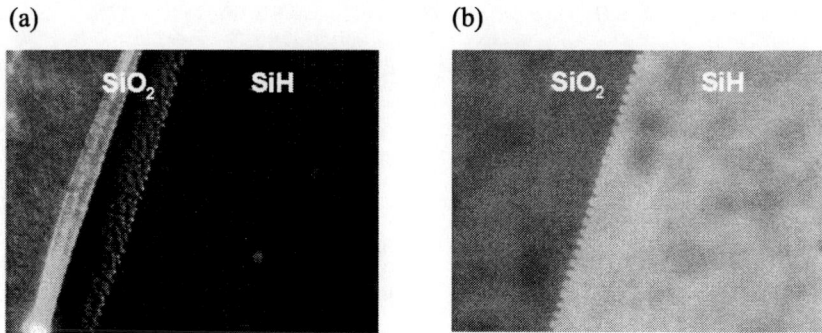

FIGURE 4. Demonstration of adsorption of DNA in SiO₂/SiH pattern. (a) Fluorescence microscope images of Poly(dG-dC)·Poly(dG-dC) self- assemble structure stained by Yo-Pro (1 μM) on SiO₂/SiH pattern. (×2000). (b) Optical microscope image of Poly(dG-dC)·Poly(dG-dC) self- assemble structure stained by Yo-Pro (1 μM) on SiO₂/SiH pattern. (×2000). The observation by fluorescence (a) and optical microscope (b) were performed in the same site.

4.   Summary

A DNA molecule can be adsorbed on Si substrate by adding $MgCl_2$ to DNA solution. In addition, DNA molecule is selectively adsorbed to SiO_2 surface in the SiO_2/SiH pattern and DNA patterning is made on the Si substrate by using DNA solution containing 0.1 mM $MgCl_2$. Since the DNA molecule and the SiO_2 surface are electrostatically combined through Mg^{2+}, the adsorption of the DNA molecule in the SiO_2/SiH pattern is depended on the concentration of $MgCl_2$ and the difference of the chemical property between the SiO_2 surface and the SiH surface. The development of a DNA device combined with silicon technology can be expected by applying this patterning.

ACKNOWLEDGMENTS

This work was supported by the Center of Excellence (COE) program under the Ministry of Education, Science, Sports and Culture of Japan and Japan Science and Techonology Corpolation (JST).

REFERENCES

1. Kurihara, K., Iwadate, K., Namatsu, H., Nagase, M., and Murase, K., *J. Vac. Sci. Technol. B* **13**, 2170-2174 (1995).
2. Carr, D. W., and Craighead, H. G., *J. Vac. Sci. Technol. B* **15**, 2760-2763 (1997).
3. Okahata, Y., Kobayashi, T., Tanaka, K., and Shimomura, M., *J. Am. Chem. Soc.* **120**, 6165-6166 (1998).
4. Tran, P., Alavi, B., and Gruner, G., *Phys. Rev. Lett.* **85**, 1564-1567 (2000).

5. Pablo, P. J., Moreno-Herrero, F., Colchero, J., Herrero, J. G., Baró, P., Herrero, A. M., Ordejón, P., Soler, J. M., and Artacho, E., *Phys. Rev. Lett.* **85**, 4992-4995 (2000).
6. Porath, D., Bezryadln, A., Vries, S., and Dekker, C., *Nature* **403**, 635-638 (2000).
7. Nakayama, H., Ohno, H., and Okahata, Y., *Chem. Commun.* 2300-2301 (2001).
8. Watanabe, H., Manabe, C., Shigematsu, T., Shimatani, K., and Shimizu, M., *Appl. Phys. Lett.* **79**, 2462-2464 (2001).
9. Storm, A. J., Noort, J., Vries, S., and Dekker, C., *Appl. Phys. Lett.* **79**, 3881-3883 (2001).
10. Hwang, J. S., Kong, K. J., Ahn, D., Lee, G. S., Ahn, D. J., and Hwang, S. W., *Appl. Phys. Lett.* **81**, 1134-1136 (2002).
11. Zhang, Y., Austin, R. H., Kraeft, J., Cox, E. C., and Ong, N. P., *Phys. Rev. Lett.* **89**, 198102 (2002).
12. Bockrath, M., Markovic, N., Shepard, A., Tinkham, M., Gurevich, L., Kouwenhoven, L. P., Wu, M. W., and Sohn, L. L., *Nano Lett.* **2**, 187-190 (2002).
13. Rakitin, A., Aich, P., Papadopoulos, C., Kobzar, Y., Vedeneev, A. S., Lee, J. S., and Xu, J. M., *Phys. Rev. Lett.* **86**, 3670-3673 (2001).
14. Hjort, M., and Staftröm, S., *Phys. Rev. Lett.* **87**, 228101 (2001).
15. Felice, R. D., Calzolari, A., and Molinari, E., *Phys. Rev. B* **65**, 45104 (2001).
16. Shigematsu, T., Shimotani, K., Chikira, M., Watanabe, H., and Shimizu, M., *J. Chem. Phys.* **118**, 4245-4252 (2003).
17. Kutnjak, Z., and Filipi_, C., *Phys. Rev. Lett.* **90**, 98101 (2003).
18. Adessi, C., and Anantram, M. P., *Appl. Phys. Lett.* **82**, 2353-2355 (2002).
19. Hartzell, B., McCord, B., Asare, D., Chen, H., Heremans, J. J., and Soghomonian, V., *Appl. Phys. Lett.* **82**, 4800-4802 (2003).
20. Lei, C. H., Das, A., Elliott, M., and Macdonald, J. E., *Appl. Phys. Lett.* **83**, 482-484 (2003).
21. Taniguchi, M., Lee, H-Y., Tanaka, H., and Kawai, T., *Jpn. J. Appl. Phys.* **42**, pp. L215-L216 (2003).
22. Kanno, T., Tanaka, H., Miyoshi, N., and Kawai, T., *Jpn. J. Appl. Phys.* **39**, pp.L269-L270. (2000).
23. Tanaka, S., Maeda, Y., Cai, L.-T., Tabata, H., and Kawai, T., *Jpn. J. Appl. Phys.* **40**, pp.4217-4220 (2001).
24. Tanaka, S., Cai, L.-T., Tabata, H., and Kawai, T., *Jpn. J. Appl. Phys.* **40**, pp.L407-L409 (2001).
25 Tanii, T., Ishibashi, K., Ohta, K., Hara, K., and Ohdomari, I., *Ext. Abstr. (47th Spring Meet. 2000); Japan Society of Applied Physics and Related Societies,* 29a-A-6.
26. Yasuda, T., Ma, Y., Chen, Y. L., Lucovsky, G., and Maher, D., *J. Vac. Sci. & Technol. A* **11**, 945-951 (1993).
27. Thundat, T., Allison, D. P., Warmack, R. J., Doktycz, M. J., Jacobson, K. B., and Brown, G. M., *J. Vac. Sci. & Technol. A* **11**, 824-828 (1993).
28. Rabke, C. E., Wenzler, L. A., and Beebe, Jr., T. P., *Scanning Microscopy* **8**, 471-480 (1994).
29. Muir, T., Morales, E., Root, J., Kumar, I., Garcia, B., Vellandi, C., Jenigian, D., Marsh, T., Henderson, E., and Vesenka, J., *J. Vac. Sci. & Technol. A* **16**, 1172-1177 (1998).
30. Zheng, J., Li, Z., Wu, A., and Zhou, H., *Biophysical Chemistry* **104**, 37-43 (2003).
31. Ouyang, Z.-Q., Hu, J., Chaen, S.-F., Sun, J.-L., and Li, M.-Q., *J. Vac. Sci. & Technol. B* **15**, 1385-1387 (1997).

Multi-Level Self Organization Process For A Parallel Fabrication Of Aligned Metal Structures In Microelectrode Gaps Using DNA And Metal Nanoparticles

Wolfgang Fritzsche, Gunter Maubach, Andrea Csaki, Detlef Born, Uwe Klenz

Institute for Physical High Technology Jena, Germany

Abstract. A fabrication scheme for the generation of metal nanostructures integrated in microelectrode gap arrays has been developed. The scheme uses self-organization of molecular units such as long DNA and metal nanoparticles based on specific interactions. Thereby, it is open for parallelization as a typical requirement for future application of this approach. The assembly process is explained and demonstrated, and the results of ultramicroscopic characterization is presented.

INTRODUCTION

Molecular self assembly based on specific binding pairs is a promising approach in DNA-based nanoelectronics. So DNA binds to prestructured microelectrodes or metal nanoparticles, and metal layers are deposited along stretched DNA. However, to transform these experiments into a technology, an optimal integration with today's technical platforms is required. This integration is usually realized by low-yield random adsorption onto prestructured electrode arrangements or by physical nanotechniques, such as e-beam lithography or conductive AFM. These serial approaches are time-consuming and limit the number of possible experiments.

An ideal process would be highly parallel in order to realize high throughput for both further experimental developments and possible future applications. We proposed an approach based on microstructured substrates and long DNA molecules that bind as individual molecules in an extended conformation at defined positions, and can be used for positioning of e.g. nanoparticles in further steps [1]. Utilizing micrometer-sized electrode structures and 16 μm-long DNA (double-stranded and linear), a self-assembly of individual molecules on each electrode was demonstrated [2]. However, the required chemical functionalization of the electrodes and the need for a flow-chamber complicated the experiments. Utilizing the receding meniscus of a drying droplet for the extension of immobilized DNA [3] and overlaying this process with microstructured electrode arrays, an alignment of the DNA along the electrodes could be achieved [4].

CP725, *DNA-Based Molecular Electronics: International Symposium*, edited by W. Fritzsche
© 2004 American Institute of Physics 0-7354-0206-X/04/$22.00

Here we report about further steps towards nanoelectronic demonstrators: The prepared microelectrode arrays with gaps containing one DNA structure each spanning the gap were the starting point for more levels of self-organization by positioning metal nanoparticles along the stretched DNA in a specific way.

MATERIALS & METHODS

Microelectrode preparation and substrate functionalization is explained elsewhere [4]. Briefly, lithographic standard lift-off processes were used to fabricate gold electrode arrays on silicon oxide substrates with electrode widths and gap sizes in the lower micrometer range. DNA of the lambda bacteriophage (lambda DNA) in a concentration of 250 ng/µl was 500 and 100fold diluted for the experiments [4].

Gold nanoparticles in the 20 nm diameter range with a high affinity to DNA were used [4]. The DNA is immobilized prior to the nanoparticle incubation.

Fluorescence labeling was with YOYO-1 (Molecular Probes), and detected using a microscope Axiotech (Carl Zeiss, Jena, Germany) equipped with a CCD camera Sensicam (PCO Computer Optics, Kehlheim, Germany). A scanning force microscope NanoScope II with a measurement head Dimension 3100 (Digital Instruments, Santa Barbara, CA) was used in tapping mode for AFM imaging. Scanning electron microscopy (SEM) was performed with a JSM 6700 F (Jeol, Zaventem, Belgien). The DNA was therefore labeled with nanoparticles.

RESULTS & DISCUSSIONS

Realization of electrode gaps with single DNA structures in an extended and wired state is a prerequisite for studies of DNA application for molecular electronics, either using DNA direct as functional molecule after chemical modification or as base for a framework with specific binding sites. In order to fabricate such a structure, we combined methods for the adsorption of stretched DNA molecule with microstructured surfaces.

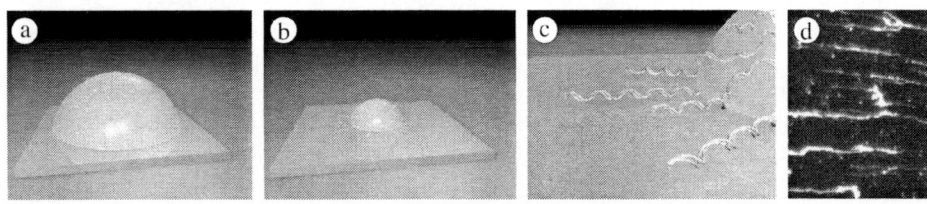

FIGURE 1: Alignment of DNA molecules due to surface-adsorption and flow induced by a receding mensicus. Molecules adsorb at one point on a functionalized substrate (a) and get aligned due to the force of the moving air-water interface (b), resulting in radially aligned immobilized molceules (c). A fluorescence image shows DNA structures (bright) immobilized in an aligned state (d).

A widely applied method for stretching DNA is based on the receding meniscus of an air-water interface, either as a droplet or on a substrate removed from a liquid reservoir comparable a Langmuir trough setup [2,5]. Thereby, surfaces with a certain affinity to DNA (e.g. positively charged) are used. When a molecule adsorbs in one point (preferentially at one end) to the surface, the liquid flow induced by the receding meniscus leads to an alignment in this direction (Fig. 1b,c). A typical result is shown in Fig. 1d, the fluorescently labeled DNA appears as bright features that are preferentially aligned in horizontal direction.

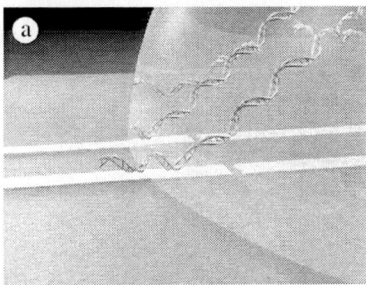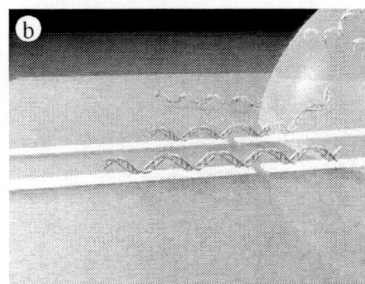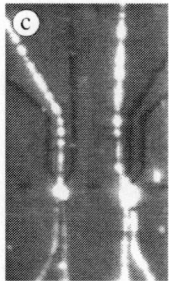

Figure 2: Positioning of DNA along electrodes by combination of receding droplet and aligned microelectrodes. The drying droplet leads to an orientation of surface-bound molecules (a), and this orientation follows the electrodes including gaps (b). c) Fluorescence image showing the DNA (bright) follwoing the metal electrode structures (dark).

This alignment was now combined with prestructured electrodes in order to achieve controlled positioning of the DNA along the electrodes and thereby also in the electrode gaps. Therefore, the droplet boundary and the orientation of the electrodes had to be aligned to be parallel, and the electrodes should exhibit a certain affinity to the DNA. The process is described in Figs. 2a and b: Extended molecules fixed in one point (one end) at the electrodes are aligned by the receding meniscus, and this alignment follows the orientation of the electrodes. If a gap occurs, the molecules will span it, resulting in an arrangement of a gap bridged by a molecular structure. An experimental result is given in Fig. 2c. DNA, visible as bright structure in the fluorescence contrast, follows the predefined gold electrode structures (dark). Two gaps (similar to the one shown in Fig. 3d) are situated in the bright spots. The enhanced fluorescence signal here is probably based on minimized quenching in the gap, because the gold electrodes usually quench (and thereby reduce) the fluorescence signal. As result, individual DNA structures are positioned by self-organization at the desired position and with the intended orientation [4].

In order to achieve electronic functionality, conductive material in defined arrangements have to be included. Metal nanoparticles could be the material of choice, due to their interesting electronic properties and because they can be functionalized to be compatible with directed binding to DNA-based structures. We investigated the use of gold nanoparticles to create structures along the DNA aligned in the electrode gaps. The basic idea is described in the schemes in Fig. 3a and b: Positioned DNA serves as binding site for functionalized metal nanoparticles. Experiments with DNA molecules on glass substrates without electrodes demonstrated the specificity of the nanoparticle binding and the low background due to unspecific binding (Fig. 3c).

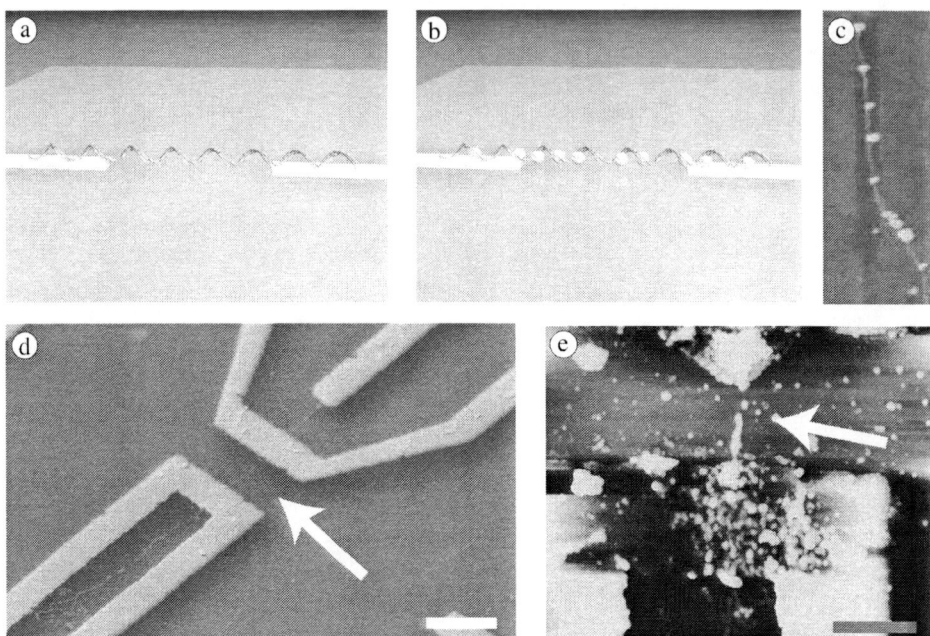

FIGURE 3: Nanoparticle binding onto DNA in electrode gaps. DNA positioned in nanoelectrode gaps (a) is functionalized by nanoparticles that bind specifically to it due to electrostatic attraction (b). c) Nanoparticles (about 20 nm diamter) positioned along an individual DNA molecule. d) SEM image of nanoparticles bound along a DNA structure positioned in an electrode gap (arrow). Bar 2μm. e) The same gap as shown in d) imaged by AFM. Bar 1μm.

So the nanoparticles were applied to chip substrates with DNA positioned in electrode gaps. Observations by SEM yielded thread-like features spanning the gap (Fig. 3d), that were identified as modified DNA based on comparative DNA imaging before and after particle binding (data not shown). The presence of a contrast in the SEM points to metal particles along the DNA. However, a closer examination of the images revealed a inhomogeneity of the DNA structure, pointing to an uneven distribution of the particles along the molecule. In order to characterize this feature as a key parameter for possible electronic applications, AFM imaging was conducted. A zoom of the gap from Fig. 3d is given in Fig. 3e. The topographic contrast reveals a certain level of unspecific binding of nanoparticles in the gap region, but clearly below the possibility for competing conductive pathways between the electrodes. The DNA structure in the gap (arrow) is only partially visible, some regions are apparently without attached particles. The large structure protruding from the upper electrode is assumed to be of nonconductive character, because it does not predominantly show up in the SEM image (Fig. 3d).

The two gaps in the metallized DNA are of approximately 50 and 100 nm width, respectively. So these values are much to large for tunneling processes. Electric characterization of this and similar gaps yielded resistances in the range of 70-700 GOhm [6]. Based on control experiments excluding artifacts due to contributions by adsorption layers on the whole substrate etc, we explain the measurement by a minimal conductivity along the adsorbed DNA which is probably embedded in salt or other residues.

Conclusions

We could demonstrate a novel approach for the fabrication of metal nanostructures towards individual molecular structures, based on self-organization and therefore potentially open for a high parallelization as required for extended research in this field and possible future applications. An important advantage regarding other typical approaches in this field is the inherent integration into a macroscopic technical setup by means of the microelectrodes and their wiring.

Acknowledgements

We thank F. Jahn for SEM imaging. This work was funded by the Volkswagen Foundation (Priority area: Physics, Chemistry and Biology with Single Molecules) and the DFG (FR 1348/3-4).

REFERENCES

1. Fritzsche, W.; Maubach, G.; Born, D.; Köhler, J. M.; and Csaki, A. (2002) in DNA-Based Molecular Construction (Fritzsche, W., Ed.) pp 83-92, AIP Conference Proceedings 640.
2. Maubach, G.; Csaki, A.; Seidel, R.; Mertig, M.; Pompe, W.; Born, D.; and Fritzsche, W. Nanotechnology 2003, 14, 546-550.
3. Bensimon, A.; Simon, A.; Chiffaudel, A.; Croquette, V.; Heslot, F.; and Bensimon, D. Science 1994, 265, 2096-2098.
4. Maubach, G.; and Fritzsche, W. Nano Letters 2004, 4, 607-611.
5. Hu, J.; Zhang, Y.; Gao, H.; Li, M.; and Hartmann, U. Nanoletters 2002,2, 55-57.
6. Maubach, G.; Csaki, A.; Born, D.; and Fritzsche, W. 2004, in preparation.

DNA DERIVATIZATION

Non-Covalent Binding of DNA to Carbon Nanotubes Controlled by Biological Recognition Complex

Laurence Goux-Capes[1]*, Arianna Filoramo[1], Denis Cote[2], Emmanuel Valentin[3], Jean-Philippe Bourgoin[1], Jean-Nöel Patillon[3]

1-Laboratoire d'Electronique Moléculaire CEA Saclay, DSM/DRECAM/SCM, 91191 Gif-sur-Yvette, France
2-LPMC, Ecole Normale Supérieure, 24 rue Lhomond, 75231 Paris Cedex 05 France
3-Centre de Recherche MOTOROLA Paris, MOTOROLA Labs Les algorythmes, 91193 Gif-sur-Yvette Cedex, France
capes@drecam.saclay.cea.fr

Abstract. Single wall carbon nanotubes (SWNTs) occupy a special place within molecular electronics. Indeed, they exist as semiconducting or metallic wires and have been used to demonstrate molecular devices like transistors, diodes or SET (single electron transistor). However, the future of this class of SWNT-based devices is strictly related to the development of a bottom-up self-assembly technique. The exceptional recognition properties of DNA molecule make it an ideal candidate for this task. Here, we describe a non-covalent method to connect carbon nanotubes to DNA strands using the streptavidin/biotin complex. Control experiments show that in absence of biotin, the DNA strand do not bind to SWNT. The binding of SWNT to DNA strand has also been carefully checked by washing experiments, showing the strength of the DNA anchorage on SWNTs. Combining this approach with molecular combing enable us to align nanotubes on substrate.

INTRODUCTION

Molecular Electronics is increasingly studied as a candidate alternative technology to CMOS for two main reasons. First, it inherently deals with the size of molecular objects, which is foreseen as a possible answer to the miniaturization problem. Second, it is a natural field for the use of self-assembling techniques, supposedly the best way to reduce the fabrication costs. Among the molecular objects, single wall carbon nanotubes (SWNTs) have been intensively studied and recently a sudden acceleration of the field occurred with the demonstration of room temperature single electron transistor (SET) [1], and of SWNT transistors showing gain above unity [2]. This result was immediately applied to the realization of logical gates mimicking the CMOS ones but with a lateral channel extension reduced to 1 nm [3]. However, all these demonstrations rely on randomly deposited SWNTs and are not really suitable within the bottom-up philosophy of molecular electronics. To fully exploit the unique properties of SWNT devices it is very desirable to be able to organize them by means of self- and possibly directed-assembly techniques.

CP725, *DNA-Based Molecular Electronics: International Symposium*, edited by W. Fritzsche
© 2004 American Institute of Physics 0-7354-0206-X/04/$22.00

RESULTS and DISCUSSION

To solve the SWNTs random deposition issue two post-synthesis method can be drafted. The first one is to achieve a selective placement of SWNTs on predefined region of substrate. Our lab was pioneer in the use of self-assembled monolayers (SAMs), which modify the surface properties of a prepatterned substrate [4]. We obtained by this way the suitable high densities of SWNTs deposition necessary for the realisation of integrated devices [5] (Fig.1).

Such positioned SWNTs have been electrically contacted to realize high performance transistors, which very well compare with state-of-the-art carbon nanotubes field effect transistors [6]. However, in this approach the patterns and the electrodes are realized by standard lithographic techniques. A real technological breakthrough would be to develop a complete molecular scale bottom-up method.

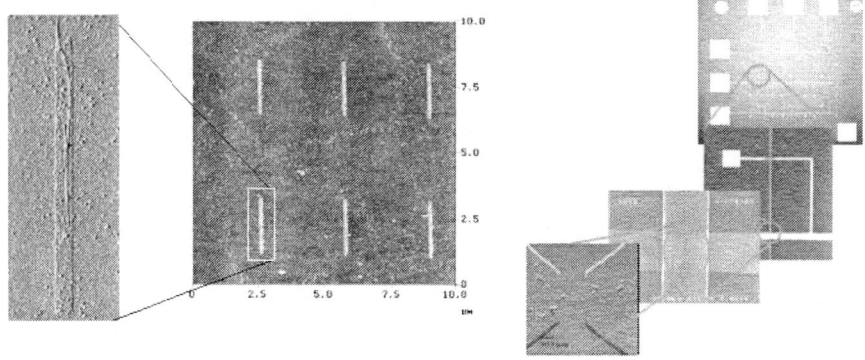

FIGURE 1. Array of SWNTs transistors. Left: SWNTs selectively deposited on APTS patch patterned on SiO$_2$. Right: lithography steps to connect SWNTs to electrodes

The second approach could solve this challenge using DNA scaffold to realize a site-controlled implementation of nanocomponents. Indeed the unique intra- and intermolecular recognition properties of DNA has been already used to build-up scaffold structures and position nanoparticles [7].

Among the several key points required to perform SWNTs devices network assembled on DNA scaffold, one of them is to control SWNT-DNA binding. Most results previously reported on this topic deal with covalent chemistry based on carboxylic acid defects groups present on SWNTs sidewall [8-12]. Indeed, the oxidation step performed during the usual purification process of the SWNTs produces such defects in relatively low amounts (estimated to about 2-3%) [13,14]. To improve the reaction yield of covalent methods, strong acidic treatments are generally required to increase the number of defects on the nanotube walls. The effect of introducing a large number of defects along the SWNT sidewalls is still not clearly known but is believed to strongly affect their electronic properties. Indeed it is known that the electronic properties of nanotubes are affected by their environment and this feature has been used for sensor applications [15] or in particular conditions for transistor performances enhancing [16]. Here, we describe a non-covalent method to link carbon nanotubes to DNA strands. This avoids the need of any additional aggressive acid treatment, and preserves the SWNTs in their original (mainly sp^2) structure.

Different works have been reported on the association of biological molecules (DNA and/or proteins) with nanotubes by means of non-covalent chemistry. A strong interaction of DNA with nanotube surfaces has been suggested [17,18] and the attachment of streptavidin and other proteins to the nanotube was reported too [19]. In this line, we studied the attachment conditions between DNA and SWNT and succeeded to achieve the control of this process by the use of streptavidin/biotin complex.

The streptavidin is a relatively small protein (60,000 Dalton) composed of four identical subunits. The mechanism that binds streptavidin to a sp^2 nanotube surface is probably related to hydrophobic interaction [19]. Indeed, this molecule is known to bind to hydrophobic surface [20]. In our studies, we first tested the interaction between SWNTs and streptavidin by achieving a complete coating of the nanotube surface by this protein. In addition, we observed that this effective hydrophobic interaction provides a simple way to functionalise SWNTs with nanoparticles. Indeed, incubation of streptavidin labeled gold nanoparticles with an aqueous solution of SWNTs allowed us to attach the gold nanoparticles to the SWNTs sidewalls (Fig 2 c).

The streptavidin protein is particularly well studied for its various biochemical applications because of its high affinity to biotin. Indeed, each of the streptavidin subunits has an active binding site for biotin molecule and the streptavidin/biotin system has one of the largest free energies of association yet observed for non-covalent binding of a protein and small ligand in aqueous solution (K_assoc = 10^{14} M^{-1}). Moreover, these complexes are also extremely stable over a wide range of temperature and pH.

To perform attachment of SWNTs to DNA, the streptavidin coated nanotubes were incubated with 10 kb 5'-biotin-DNA double strands. The characterization was performed by means of AFM imaging of the deposited products of reaction on 3_aminopropyltriethoxysilane (APTS) layer deposited on silicon wafer or mica substrate. The AFM image shown in Fig.2a clearly demonstrates the attachment of a DNA strand to a SWNT. The height of the coated SWNT is 4 nm. In order to discriminate whether the DNA is bound to single SWNT or to a bundle of SWNTs, one needs to know the height of a single streptavidin monolayer. This height has been already discussed in literature. Values in the 3-5 nm range were reported [13; 21-25]. We also performed such measurements of the streptavidin monolayer height in our experimental conditions and found a height of roughly 3 nm. Based on this value, we conclude that our method permits to connect a single SWNT (or a small bundle of very few nanotubes) to the DNA strand.

Control experiments have been performed with non-biotinylated DNA strands. In this case, the streptavidin coating of SWNT was effective but no connection between SWNT and DNA was observed, as reported in Fig. 2b.

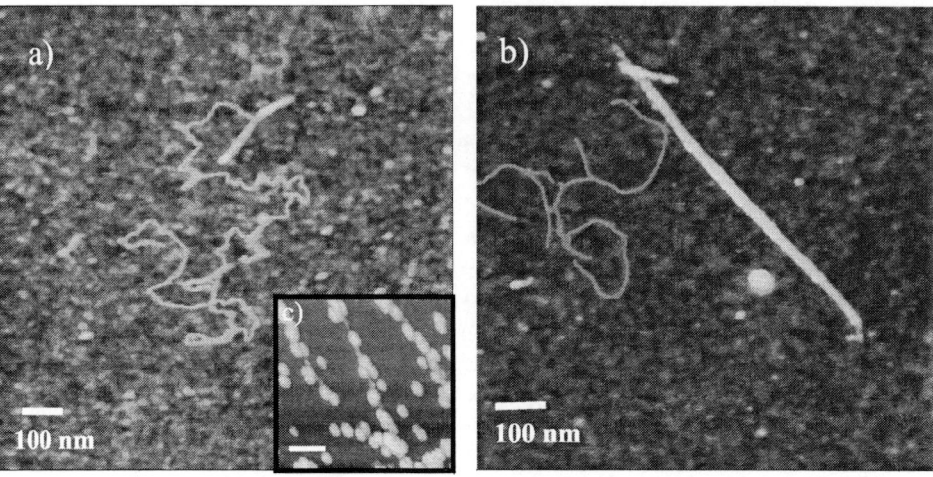

FIGURE 2. (2a) Attachment of a DNA strand to SWNT. (2b) Control experiment. (2c) Streptavidin gold nanoparticle coated nanotubes. AFM image in Fig.2a presents a typical result of the attachment of a DNA strand to a SWNT. In Fig.2b a control experiment with non-biotinylated DNA is reported. In this case, the streptavidin coating of SWNT was noted but no connection between SWNT and DNA was observed. [DNA strands have been colored for clarity. Original images are available upon request].

Moreover, "washing" experiments were performed in order to verify and test the achieved anchorage of the DNA strands to SWNT. Indeed, by carefully adjusting the rinsing conditions, it is possible to allow DNA mobility on the surface while SWNTs do not move. The APTS substrate on which the DNA-SWNT complex has been preliminary deposited, located and imaged by AFM was submerged in a basic solution. This solution decreases the interaction of the DNA strand with the APTS and leaves the molecule free to move around its anchorage point on the SWNT stuck on the surface. Then the surface is dried and the DNA-SWNT target is imaged again by AFM. Fig.3 shows a DNA-SWNT complex imaged as initially deposited on the APTS surface (3a) and the same object after washing (3b). It is clear from Fig.3 that the DNA strand moves during the washing since its shape on the surface varies. Moreover, the NT remains fixed and most importantly, the anchorage of DNA to SWNT remains after washing even for conditions to a break of the DNA strand.

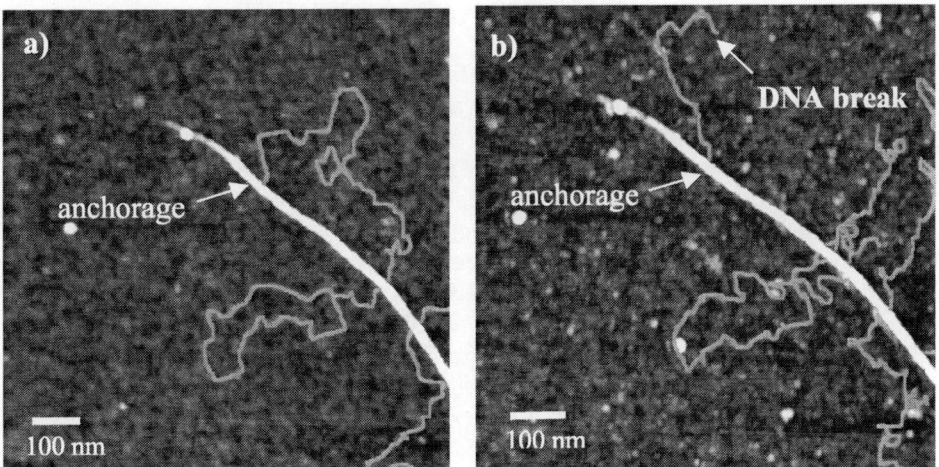

FIGURE 3. "Washing" experiment performed on a DNA-SWNTs complex (3a) AFM image of a DNA strand linked to a SWNT on a localized substrate region (before washing). (3b) AFM image of the same region with the rearranged DNA strand linked to the same SWNT after washing treatment. [DNA strands have been colored for clarity. Original images are available upon request].

Combining DNA-SWNTs binding mediated by streptavidin-biotin recognition and molecular combing we were able to align SWNTs on a substrate. Figure 4 illustrate the combing of DNA-SWNTs adducts on a silicium wafer.

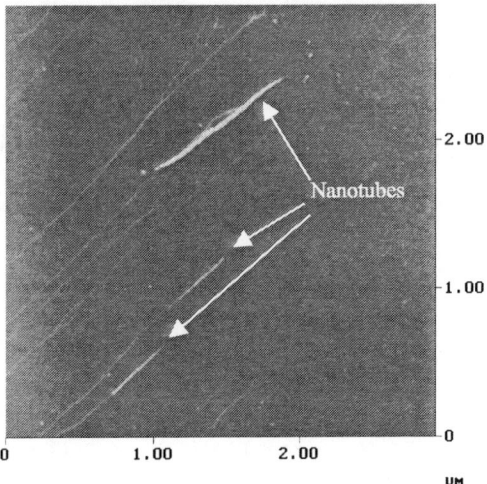

FIGURE 4. Combing of SWNTs-DNA on SiO_2 surface. The nanotubes at the end of DNA strands are indicated by arrow.

21

EXPERIMENTAL PROCEDURE

Binding SWNT to DNA by streptavidin-biotin recognition.
A double strand 10 kb 5'-Biotin-DNA was prepared by polymerase chain reaction using long expand PCR system (Roche Diagnostics) on lambda DNA with designed primers: 5'- ATACgCTgTATTCAgCAACACCgTCAggAACACg-3' and 5'- Biotin-CTgATgAgTTCgTgTCCgTACAACTggCgTAATC-3'.
SWNTs are synthesized by laser ablation (O. Jost, Dresden Institute) and purified according to a homelab two steps procedure combining nitric acid and hydrogen peroxide [26]. Purified SWNTs were dispersed in water by overnight sonication (5 10⁻² mg/ml). Streptavidin aqueous solution is added to SWNTs dispersion and mix gently during one hour at room temperature. Centrifugation was subsequently carried out to remove insoluble particles from the dispersion.
5 µl of a 6 nM solution of 10 kb 5'-Biotin -DNA is added to 20 µl of the streptavidin coated SWNTs solution. The system was thereafter incubated two hours at room temperature afterwhat it is diluted in 80 µl of Tris buffer, pH = 8.
Control experiments.
a) Reference samples: Same experiment has been performed with non-biotinylated DNA. The DNA fragments were incubated with streptavidin coated nanotubes.
b) Washing experiments: DNA-SWNT complex are deposited on APTS (3_aminopropyltriethoxysilane) layer then located and imaged by AFM. The substrate is then submerged in a basic solution during 15 min (Tris pH=9, 10 mM NaCl), rinsed in water and dried under nitrogen. Once the surface is dried, the DNA-SWNT target is imaged again by AFM.
Combing experiment: DNA stretching has been adapted from literature using a vertical translation stage [27]. The combing has been performed on cleaned silicon wafer with DNA-SWNTs solutions (MES pH = 5,5; 5 mM MgCl₂).
AFM characterization: To characterize DNA-SWNT coupling, a drop of the reaction product is deposited on an APTS layer. After 5 min of incubation, the surface is rinsed carefully with the buffer solution, then with water and dried under nitrogen. To image a particular object after successive treatments the substrate has been carefully marked by lithography techniques.

CONCLUSION

In conclusion, we demonstrated that the linkage by streptavidin/biotin complex of DNA to SWNT is extremely effective, strong and stable. In addition this approach, combined with molecular combing, allow to align SWNTs on a substrate. Moreover, the concept of non-covalent chemistry on SWNTs can be extended to other biological molecules displaying hydrophobic interactions with carbon surfaces. Indeed, for some of them the binding to nanotube sidewalls has been already reported (cytochrome *c*, ferritin, IgC Antibody …[28]). Therefore, one can envision performing DNA to SWNT connection using the recognition of other ligand-receptor systems or by direct binding of protein-DNA complexes.

The realization of DNA-functionalised SWNTs opens the route towards the formation of nanotubes arrays by DNA guided self-assembly on a DNA scaffold. Moreover, our results combined with progresses in DNA metallisation [29-31] should permit the implementation of self-assembled molecular scale electronic systems. Very recently a first device has been demonstrated [32] based on this concept. Next challenge is to realize an assembly of such devices on DNA scaffold to perform signal processing with performance at least comparable to self-assembled monolayer approach.

ACKNOWLEDGEMENTS

This work is partially supported by the European contracts IST-1999-13099 DNA-based electronics and IST-1999-10593 SATURN. The authors wish to thank U. Bockelmann (Ecole Normale Supérieure-Paris) for useful discussions.

REFERENCES

1. Postma H. C., Teepen T., Yao Z., Grifoni M. and Dekker C., *Science* **293**, 76 (2001)
2. Derycke V., Martel R., Appenzeller J. and Avouris Ph., *Nano Letters* **1**, 453 (2001)
3. Bachtold A., Hadley P., Nakanishi T. and Dekker C., *Science* **294**, 1317 (2001)
4. Choi KH, Bourgoin JP, Auvray S, Esteve D, Duesberg GS, Roth S, Burghard M, *Surface Science*, **462 (1-3)**, 195 (2000)
5. Valentin E., Auvray S., Goethals J., Lewenstein J., Capes L., Filoramo A., Ribayrol A., Bourgoin J-P, Patillon J-N, *Microelectronic Engineering* **61-2**, 491 (2002)
6. Valentin E., Auvray S., Filoramo A., Ribayrol A, Goffman M., Goethals J, Capes L., Bourgoin J-P, Patillon J-N, *Materials Research Society Symposium - Proceedings*, **772**, 2003, p 201-207; Valentin E., Auvray S., Filoramo A., Ribayrol A, Goffman M., Goethals J, Capes L., Bourgoin J-P, Patillon J-N; *NATO Sciences series: Molecular Nanowires and other Quantum Objects* –Ed. Sasha Alexandrov, Jure Demsar and Igor Yanson (2004)
7. Seeman NC, *Nature*, **421**, 427 (2003); Alivisatos AP, Johnsson KP, Peng XG, Wilson TE, Loweth CJ, Bruchez MP, Schultz PG, *Nature* **382 (6592)**, 609 (1996); C. Niemeyer, B. Ceyhan, *Angew. Chem. Int.Ed*, **40**, 3685 (2001); Mirkin C.A , *Inorg. Chem*, **39** 2258 (2000); Li H., Park S.A., Reif J.H.,LaBean T.H, Yan H., *J. Am. Chem. Soc.*, **126** 418 (2004)
8. S. Niyogi, M. A. Hamon, H. Hu, B. Zhao, P. Bhowmik, R. Sen, M. E. Itkis, R. C. Haddon, *Acc. Chem. Res.* **35**, 1105 (2002)
9. Dwyer C., Guthold M., Falvo M., Washburn S., Superfine R. and Erie D., *Nanotechnology* **13**, 601 (2002)
10. Baker S.E., Cai W., Lasseter T.L., Weidkamp K.P.and Hamers R.J, *Nano Letters* **2**, 1413 (2002)
11. Nguyen CV, Delzeit L, Cassell AM, Li J, Han J, Meyyappan M, *Nano Letters* **2**, 1079 (2002)
12. Williams K.A., Veenhuizen P.T.M., de la Torre B.G., Eritja R. and Dekker C., *Nature* **420**, 19 (2002)
13. Rinzler A.G., Liu J., Dai H., Nikolaev P., Huffman C.B., Rodríguez-Macías F.J., Boul P.J., Lu A.H., Heymann D., Colbert D.T., Lee R.S., Fisher J.E., Rao A.M., Eklund P.C., Smalley R.E., *Appl. Physics A*. **67**, 69 (1996)
14. Hu H., Bhowmik P., Zhao B., Hamon M.A., Itkis M.E., Haddon R.C., *Chem.Phys.Lett.* **345**, 25 (2001)
15. Besteman K, Lee JO, Wiertz FGM, Heering HA, Dekker C, *Nano Letters* **3 (6)** 727 (2003)
16. Auvray S., Borghetti J., Goffman M.F., Filoramo A., Derycke V., Bourgoin J.-P., *Applied Physics Letters*, **84**,25 5106 (2004)

17. Guo Z., Sadler P.J.and Tsang S.C., *Adv.Mater.* **10**, 701 (1998)
18. Ming Z., Jagota A., Semke E.D, Diner B.A, McLean R.S., Lustig S.R., Richardson R.E. and Tassin.G., *Nature Materials*, **2**, 338 (2003)
19. Balavoine F., Schultz P., Richard C, Mallouh, V.; Ebbesen, T.W.; Mioskowski, C., *Angew. Chem. Int. Ed.* **38**, 1912 (1999)
20. Furuno T., Sasabe H., *Biophys.J.* **65**, 1714 (1993)
21. Hendrickson W.A., Pahler A., Smith J.L., Satow Y., Merrit E.A. and Phizackerley R.P., *Proc. Natl. Acad. Sci. USA* **86**, 2190 (1989)
22. Scheuring S., Muller D.J., Ringler P., Heymann J.B. and Engel A., *Journal of Microscopy* **193**, 28 (1999)
23. Darst S.A., Ahlers M., Meller P.H., Kubalek E.W., Blankenburg R., Ribi H.O., Ringsdorf H. and Kornberg R.D., *Biophys. J.* **59**, 387 (1991)
24. Gau J.J., Lan E.H., Dunn B., Ho C.M. and Woo J.C.S, *Biosensors and Bioelectronics* **16**, 745 (2001)
25. Fritzsche W., Ermantraut E. and Kohler J.M., *Scanning* **20**, 106 (1998)
26. Capes L., Valentin E., Esnouf S., Ribayrol A., Jost O., Filoramo A., Patillon J.-N., *Nanotechnology, 2002. IEEE-NANO 2002,* 439
27. Allemand, J. F.; Bensimon, D.; Jullien, L.; Bensimon, A.; Croquette,V. *Biophys. J.*, **4** 73 (1997) 2064.
28. Erlanger B., Chen B. X., Zhu M., Brus L., *Nanoletters*, **1**, 465 (2001); Azamian B.R., Davis J.J., Coleman K.S., Bagshaw C.B., Green M.L.H., *J. Am. Chem. Soc.* **124**, 12664 (2002)
29. Keren K., Krueger M., Gilad R., Ben-Yoseph G., Sivan U. and Braun E., *Science* **297**, 72 (2002); Keren K, Berman R.S, Braun E., *Nanoletter* **4**, 323 (2004)
30. Richter J., Miertig M., Pompe W., Mönch I., Schackert H.K., *Appl. Phys. Lett.* **78**, 536 (2001)
31. Harnack O., Ford W. E., Yasuda A., Wessels J. M., *Nanolett.* **2**, 919 (2002)
32. Keren K, Berman RS, Buchstab E, Sivan U, Braun E, Science, **302**, 1380 (2003)

Self-Assembly Experiments with PNA-Derivatized Carbon Nanotubes

Remco den Dulk, Keith A. Williams, Peter T.M. Veenhuizen,
Martijn C. de Koning*, Mark Overhand*, Cees Dekker

*Molecular Biophysics Group, Kavli Institute of Nanoscience Delft,
Delft University of Technology, The Netherlands*
**Leiden Institute of Chemistry, Gorlaeus Laboratories,
University of Leiden, The Netherlands*

Abstract. We are conducting experiments to fabricate nanotube-based field effect transistors (FETs) using the molecular recognition properties of DNA. For this purpose, we have prepared single-walled carbon nanotubes derivatized with PNA (peptide nucleic acid, a DNA analog) and have studied their attachment to free, complementary DNA. We are currently examining the prospects for assembling devices by hybridization of the PNA-labeled nanotubes to DNA-functionalized electrodes.

Introduction

Single-walled carbon nanotubes (SWNT) combine remarkable electronic and chemical properties that make them ideal molecular channels in field-effect and single-electron transistors [1-3]. Since the construction of the first prototype nanotube-based devices, numerous research groups have worked toward ever more complex designs. However, the precise positioning of nanotubes is difficult and locating them one at a time is inefficient. We are developing a bottom-up, self-assembly strategy for integrating nanotubes into electronic circuits in a massively parallel fashion. We have covalently derivatized SWNT with peptide nucleic acid (PNA, a DNA analog with a chemically robust, uncharged, pseudo-peptide backbone [4]), and are working toward a PNA-DNA hybridization scheme for placement of labeled nanotubes between source and drain electrodes based on Watson-Crick, base-pairing molecular recognition. In our strategy, oligonucleotide sequences control the interactions of the PNA-SWNT macromolecule with sequence-labeled substrate features. This strategy offers the prospect of bottom-up device assembly and also represents a route to reconfigurable electronics in which geometrical operations can be performed on the coordinating DNA by enzymes or local thermal sources.

CP725, *DNA-Based Molecular Electronics: International Symposium*, edited by W. Fritzsche
© 2004 American Institute of Physics 0-7354-0206-X/04/$22.00

PNA-Derivatized Carbon Nanotubes

We have chosen to derivatize SWNT with PNA for several reasons; first, because its backbone is chemically more robust than that of DNA, PNA can be handled in convenient organic solvents such as dimethylformamide (DMF); moreover, PNA does not suffer enzymatic degradation. Second, PNA is uncharged, so that nonspecific-electrostatic interactions can be avoided. Perhaps most importantly for our purposes, the melting temperature (T_m) of a typical DNA-PNA duplex is substantially higher than its DNA-DNA counterpart, so that we can use shorter "sticky" sequences and avoid nonspecific wrapping of the SWNT by ssDNA [5].

In order to derivatize SWNT with PNA, we first form carboxylic acid groups on the SWNT by sonication in concentrated sulfuric and nitric acid. The duration of this treatment determines the length of the nanotubes. After sonication, a small amount of hydrochloric acid is added to lower the pH and ensure that the carboxylic groups are protonated. Next, the COOH-SWNT material is reacted with a mixture of 2 mM EDC (1-ethyl-3-(3-dimethylaminopropyl)carbodiimide HCl, Pierce) and 5 mM NHS (N-hydroxysuccinimide, Pierce) in dry DMF in order to form NHS esters at the carboxyl sites. Finally, peptide nucleic acid PNA bearing a solubility linker and an N-terminal amine is reacted in excess with the NHS-SWNT, resulting in the covalent, amide linkage depicted in Fig. 1. Unreacted PNA is removed by washing with 10% aqueous TFA, and the material is then dispersed in aqueous Triton-X 405 by sonication, resulting in mostly individual, PNA-derivatized SWNT.

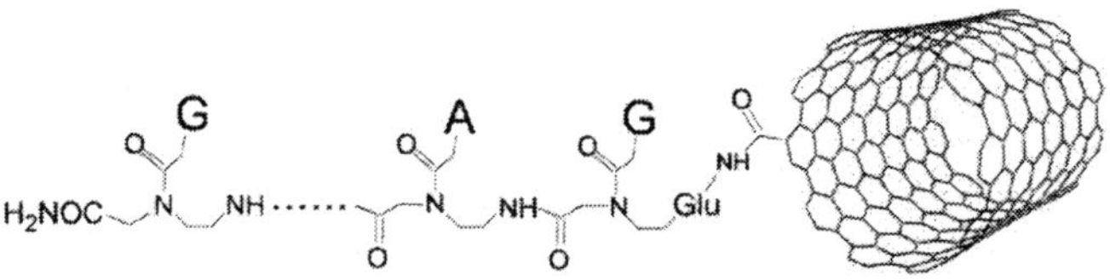

FIGURE 1. A PNA-derivatized single-walled carbon nanotube. Note the N-terminal glutamic acid spacer, which is necessary to improve the aqueous solubility of the PNA. In our work, the PNA oligonucleotide is 12 bases long. Not shown is the Triton-X 405 micelle needed to solubilize the nanotube in water and combat nonspecific interactions.

Hybridization of PNA-SWNT to DNA

To verify the hybridization of DNA to the PNA-SWNT, we used 650 bp dsDNA with a 12 base overhang at one side. This 'sticky end' (sequence: 5'-CAC-CAT-GAG-CAC-3') is complementary to the sequence of the adduct PNA. The DNA fragments were synthesized by PCR, followed by cutting with the restriction enzymes *HindIII* and *EcoR1*. Next, the 12 bp sticky ends were ligated to the DNA fragments.

PNA-SWNT dispersed in aqueous Triton-X 405 and dsDNA were mixed in the presence of 2.5 mM MgCl$_2$. The mixture was incubated for three hours at room temperature. Next, the product solution was pipetted onto freshly cleaved mica, rinsed with milli-Q water and dried with nitrogen. Ambient tapping-mode AFM images were then recorded (Fig. 2). Unlike previous samples made from roped (bundled) nanotubes grown by pulsed laser vaporization[6], the nanotubes samples used for this work were produced by the "HiPCO" process (high pressure carbonyl, carried out in the research group of Prof. R. Smalley, Rice University [7]) and pre-purified; this material consists of mostly individual nanotubes or very small ropes. Perhaps because of the purification or the lack of protection of the side-walls from neighboring nanotubes, we find that in these samples the DNA attaches not only at the ends of the nanotubes but sometimes also along the sides. We assert that this attachment is due to generation of defects in the tubes during the initial purification or sonication, and this effect can be minimized through further optimization of the chemistry.

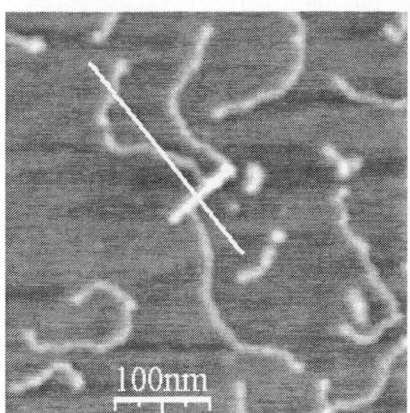 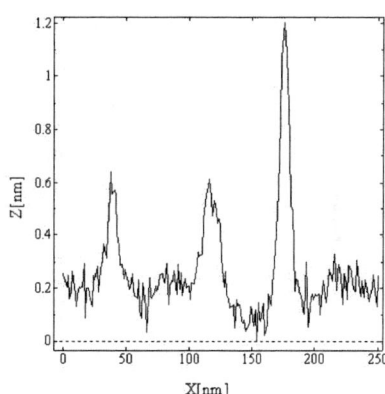

FIGURE 2. AFM height image on mica demonstrating the hybridization of doubled-stranded DNA with a 12 base pair "sticky" end to several sites on a nanotube bearing the 12 base complement. The height profile taken along the white line in the AFM image is shown on the right. The nanotube has a height of ~1.2nm; the dsDNA appears to have a height of ~0.6 nm due to tip compression.

Elements of Device Assembly

We envision the self-assembly of a PNA-SWNT device as depicted in Fig 3. In such a scheme, the PNA-SWNT macromolecules self-assemble between gold source and drain electrodes coated with complementary, thiol-bound ssDNA. This scheme requires very carefully tailored surfaces: it is essential that PNA-SWNT bind to the electrodes exclusively by means of hybridization.

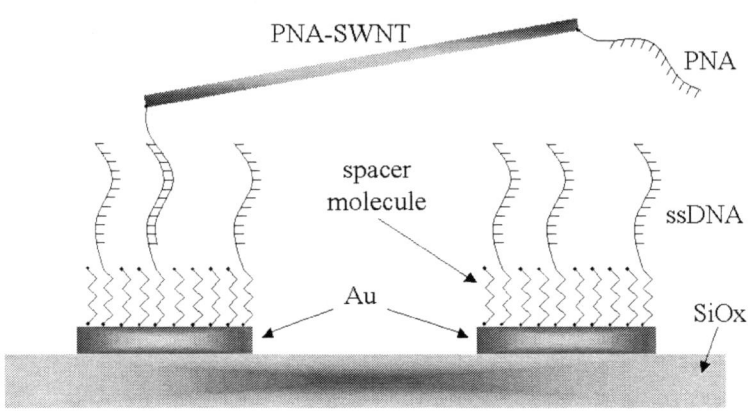

FIGURE 3. Schematic picture of the self-assembly of a plausible PNA-SWNT device.

Of course, it is necessary that the PNA-SWNT material does not bind nonspecifically. In order to achieve this, we micellize the PNA-SWNT in long-chain (~12 nm), Triton-X 405 surfactant. The gold electrode surface is made hydrophilic by covering it with a self-assembled monolayer of 3-mercaptopropionic acid (MPA, HS-$(CH_2)_2$-COOH, Sigma), which has a carboxylic group extending away from the electrode. The SiO_x is sufficiently hydrophilic due to the piranha pre-cleaning step, which generates surface Si-OH groups.

In the first step of device fabrication, Si/SiO_x substrates are patterned by electron-beam lithography and Cr/Au electrodes (3/27 nm) are formed by sequential evaporation of the metals. To remove residual resist, the samples are treated in an O_2-plasma for four minutes and subsequently sonicated in ethanol for five minutes. Just prior to the self-assembly of the alkanethiolate monolayers, the samples are cleaned with piranha (H_2O_2:H_2SO_4 ,1:4 v/v) for five minutes and subsequently rinsed with water, sonicated in ethanol for five minutes, and then rinsed again with water. Next the samples are immersed in 1_M solution of ssDNA (DMT-C_6-SS-C_6-ssDNA, Isogen Life Science) in "TEN" buffer (Tris 10mM, EDTA 1mM, NaCl 10 mM, pH 8) for 90 minutes at 20°C. The samples are rinsed with TEN and immersed in 1mM solution of MPA in TEN for 60 minutes at 20°C. The samples are rinsed again in TEN buffer and partially dried in such a way that a thin film of liquid remains on the sample. Then, 25_l of TEN containing either dsDNA or PNA-SWNT is pipetted onto a sample. The hybridization is carried out at 4°C in a sealed container to prevent the samples from completely drying. In the case of dsDNA, the required hybridization time was found to be approximately twenty hours.

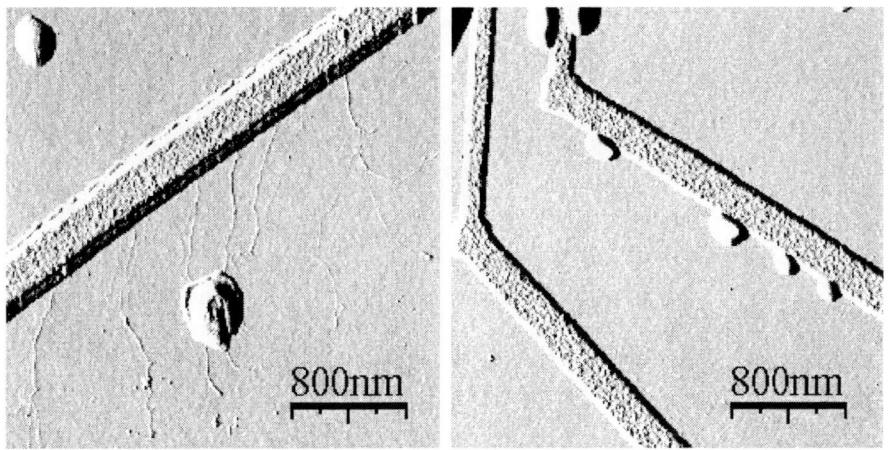

FIGURE 4. (Left) AFM image of 2.4μm 'sticky' dsDNA hybridized to complementary thiol-DNA immobilized on a lithographically defined gold electrode. (Right) Control sample, in which the target and probe DNA are not complementary. Salt crystals are also apparent in both samples and is always observed.

Self-Assembly of dsDNA to Gold Electrodes

To test the hybridization properties of the functionalized electrodes, we reacted the samples with 2.4 _m dsDNA with a 12 base 'sticky end' at one side. The dsDNA fragments were obtained from a preparation of plasmid DNA, followed by cutting with the restriction enzymes *Hind III* and *EcoR1*. The 12 base sticky ends were ligated to the dsDNA fragments. After 20 hrs of hybridization the samples were rinsed in TEN buffer at 4°C to prevent melting of the duplex (the melting temperature for this 12 base sequence is ~19°C in 10mM NaCl).

Samples were prepared with two different sequences immobilized by thiols on the electrodes. Sticky-ended dsDNA complementary to either of these two sequences was introduced, and AFM imaging was used to determine whether the dsDNA had bound to the electrodes. We observed sequence-specific binding: only electrodes bearing the complementary sequences were found to hybridize to the sticky dsDNA (see Fig. 4). Moreover, we tested dehybridization by rinsing with pure, deionized water at room temperature. This results in the complete disappearance of the dsDNA from the electrodes, again consistent with sequence-specific binding.

Attachment of PNA-SWNT to ss-DNA Labeled Gold Electrodes

We have also attempted hybridization of PNA-SWNT to electrodes bearing complementary ssDNA, but these experiments have not yet proven fruitful. Possible reasons for this include blockage by the long-chain Triton-X surfactant micelle, or simply insufficient time allowed for the hybridization at low concentration. One solution may be to use a short fragment of dsDNA with two different sticky ends as a linker. This scheme places the sequence that must hybridize to the electrode-immobilized ssDNA further outside the Triton-X micelle surrounding the nanotubes, and should thereby promote hybridization and minimize steric hindrance. We are also investigating the hybridization efficiency as a function of the PNA-SWNT concentration and the duration of the incubation.

Conclusions

In conclusion, we are working toward a scheme in which PNA-derivatized carbon nanotubes can be immobilized according to sequence-specific hybridization between source and drain electrodes. We have verified the hybridization of PNA-SWNT to complementary DNA in the presence of $MgCl_2$; we have demonstrated sequence-specific binding of "sticky" dsDNA to electrodes labeled with the sequence complement, and we have virtually eliminated nonspecific binding of the PNA-SWNT by micellizing the nanotubes with a long-chain Triton-X surfactant and by forming self-assembled monolayers of short alkanethiolates with carboxylic tails on the electrodes. Although we have not yet observed hybridization of the PNA-SWNT to the electrodes, we are now investigating the use of longer spacers on the sequences and also longer hybridization times that may be required for hybridization at low concentration.

REFERENCES

1. S.J. Tans, A.R.M. Verschueren, and C. Dekker, *Nature* **393** 49-52 (1998).
2. A. Bachtold, P. Hadley, T. Nakanishi, and C. Dekker, *Science* **294**, 1317- (2001).
3. S.J. Tans, M.H. Devoret, H. Dai, A. Thess, R.E. Smalley, L.J. Geerligs, and C. Dekker, *Nature* **386**, 474 (1997).
4. P.E. Nielsen and M. Egholm (eds.), *Peptide Nucleic Acids: Protocols and Applications*, Norfolk, England: Horizon Scientific Press, 1999, pp. 21-50.
5. M. Zheng, A. Jagota, et al., *Nature Materials* **2**, 338-342 (2003).
6. K.A. Williams, Peter T.M. Veenhuizen, Beatriz G. de la Torre, Ramon Eritja, and C. Dekker, *Nature* **420**, 761 (2002).
7. M.J. O'Connell, P. Boul, L.M. Ericson, C. Huffman, Y. Wang, E. Haroz, C. Kuper, J. Tour, K.D. Ausman, and R.E. Smalley, *Chemical Physics Letters* **342**, 265-271(2001).

Micromachined Substrates for Molecular Follow-Up in DNA-Templated Nanofabrication

Héctor A. Becerril, Allison R. Nelson[†] and Adam T. Woolley*

Department of Chemistry & Biochemistry, Brigham Young University, Provo, UT 84602, USA
[†]Current Address: Department of Physiology and Biophysics, University of California, Irvine

Abstract. We have produced micromachined silicon substrates that provide unique spatial addressing for surface-aligned DNA molecules. Repeated characterization of the same molecule using atomic force microscopy and/or scanning electron microscopy before and after nanofabrication treatments has been achieved. Utilizing these micromachined platforms as substrates in nanofabrication experiments enables the use of complementary microscopy techniques for data collection on selected features of interest at different stages of a nanofabrication process. In this way, a clear correlation of the information generated can be achieved.

INTRODUCTION

Nanoscience and nanotechnology strive to develop ways to control matter at the size scale of atoms and molecules. Important recent instrumental developments such as the scanning tunneling microscope, the atomic force microscope, and the new generations of scanning and transmission electron microscopes have allowed researchers across the globe to visualize individual atoms or molecules and to measure their physical dimensions in a manner not previously possible. These advances have provided scientists with new capabilities for manipulation and profiling of individual atoms and molecules.[1]

Expanded nanoscale imaging capabilities have proven useful in furthering the understanding of chemical processes in general, and of surface reactions in particular, where morphological or topographical changes observed by scanning probe microscopy on a given sample can often be correlated with surface chemical modifications. In some instances with favorable conditions, a complete experiment can be carried out in a microscope system to yield real-time images of surface processes. For example, atomic force microscopy (AFM) in fluid has been used to probe rheological[2-3] as well as biological[4] properties of DNA molecules on surfaces. Nevertheless, many reactions cannot be performed *in situ* within the sample cell of a scanning probe microscope, so in these cases offline treatment of the sample is the only viable alternative.

A typical offline nanofabrication experiment has the following sequence: *i)* sample preparation, *ii)* initial characterization, *iii)* chemical modification of the sample, and

CP725, DNA-Based Molecular Electronics: International Symposium, edited by W. Fritzsche
© 2004 American Institute of Physics 0-7354-0206-X/04/$22.00

iv) subsequent characterization to evaluate any effects of the treatment. Often, steps *iii)* and *iv)* are repeated multiple times on the same sample. Once the characterization data are collected, a statistical averaging method is commonly used to contrast the observations of the sample before and after the treatment(s).[5-6] A simpler and more information-rich approach would be to examine the exact same sample features both before and after offline treatment. We propose that this increased level of control over chemical processes is more in accordance with the aims of nanoscience and nanotechnology. Importantly, in this single-molecule approach the experimenter can obtain an understanding of how individual (and not necessarily equal) nanostructures on a sample respond to an applied treatment. Such information should be useful in fine-tuning experimental conditions, for example, to address one sub-group of structures on a sample while leaving the rest undisturbed.

Unfortunately, the ability to find the same nanostructure in subsequent experiments on a macroscopic substrate is not trivial. Consider the case of AFM investigation of DNA molecules. A typical AFM substrate has a surface area of about 1.0 cm^2 (10^8 μm^2), while a typical double-stranded (ds) DNA molecule may have a length of ~10 μm and a width of about 2 nm. Such a dsDNA molecule would occupy a surface area of only ~0.02 μm^2, a factor of nearly 10^{10} difference between the area of the substrate and the feature of interest! Now, suppose one wants to image the same DNA molecule after the substrate has been removed from the AFM instrument and modified in some way. Looking for that same molecule by AFM after reloading the substrate would be equivalent to searching for a 1-cm-long hair in an area greater than a football field!

The literature contains few reports where data has been obtained from a particular nanostructure and its response to consecutive offline treatments.[7-8] In general these examples follow an *a-posteriori* strategy where the experimentalist finds a nanometer-scale object of interest on the sample, and then looks for larger-size microscopic markings nearby on the surface to provide a unique identifier to enable the re-location of the feature of interest. The main disadvantage of this approach is that it does not offer a systematic means for repeated imaging of nanometer-scale features situated at any location on the surface.

The marking problem is especially challenging when one considers that to be useful, markings must be uniquely identifiable, large enough to be observed with microscopic methods, and located very close to the feature(s) of interest to facilitate finding the desired feature(s) once a microscopic mark has been located. In AFM for example, alignment marks and target structures must reside within an area encompassed by the scanner (about 200 μm x 200 μm for a scanner with sufficient resolution to observe nanometer-scale structures). Finally, the alignment marks must be sufficiently large to be identified in an optical microscope, but must also have fine enough features to be discerned in a nanometer-scale method such as AFM or electron microscopy. Thus, relying on the fortuitous finding of a suitable feature to serve as a mark is much less effective than having a rationally designed, prefabricated system of surface marking.

In this paper we describe a rational solution to the problem of repeated characterization of DNA molecules and other nanostructures on silicon substrates with AFM and scanning electron microscopy (SEM) instruments. Ours is an *a-priori* approach that produces an array of unique lithographically patterned marks on silicon substrates, which are then used for sample preparation. When a molecule or feature of interest is located, its position can be traced easily to an address on the array, and thus the feature can be found repeatedly, even utilizing different instruments.

EXPERIMENTAL

Desirable microfabricated silicon substrates must simultaneously comply with a number of criteria to be useful and general tools in repetitive feature characterization at the nanometer scale. These requirements are:

1. The whole surface of the substrate must be covered by a series of unique marks to facilitate distinct referencing of multiple features of interest on the same substrate.
2. The marks must be visible using a low-magnification optical microscope to facilitate their use for alignment of substrates in AFM and SEM.
3. The presence of these marks must not affect the preparation of the sample; in the particular case of DNA molecules, the marks must be shallow enough that they do not disturb the DNA alignment process.
4. There must be sufficient space between marks that DNA molecules or other nanostructures can be imaged on a locally undisturbed surface.
5. The substrate must be compatible with AFM and SEM. The surface roughness should be comparable to silicon or silicon dioxide substrates, and the surface should not be susceptible to charging under a SEM electron beam.

With these guidelines in mind, substrates were produced as described in the next section.

Microfabrication

Molecular localization substrates were microfabricated on p-type silicon [100] wafers. Briefly, an electronic version of the mask, containing unique 90-μm-spaced character pairs, was prepared using computer-aided design software. The pattern was transferred to a chrome-on-glass photolithographic mask using an Electromask Lasometric pattern generator. Silicon wafers were thermally oxidized in a tube furnace at 1110 °C to produce a 570 nm silicon dioxide layer, whose thickness was determined by ellipsometry. Preliminary experiments with these oxidized wafers showed that SEM imaging of nanosized structures was not feasible due to charging of the dielectric layer. However, if the silicon dioxide on the wafer surface was etched in buffered HF solution to a thickness of about 100 nm, the surface roughness was comparable to that of silicon substrates with a native oxide layer, and the SEM charging effects disappeared.

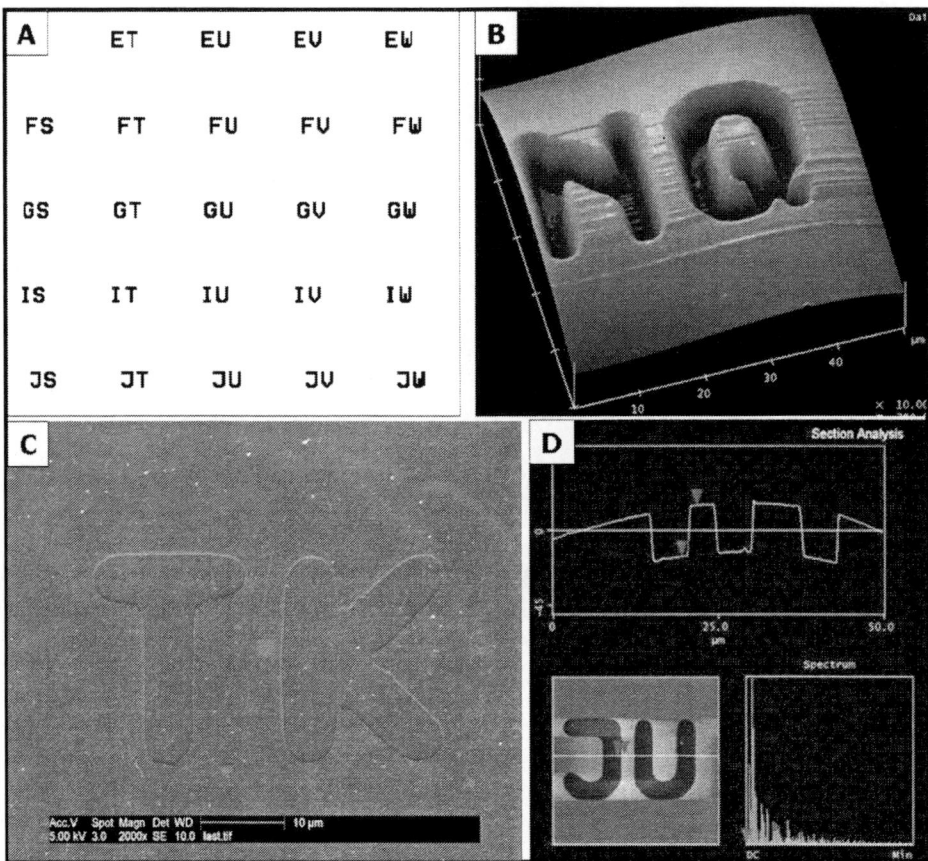

FIGURE 1. Microfabricated molecular localization substrates. A) Portion of the design file showing unique character pairs evenly spaced by 90 µm in all directions. B) Tapping-mode AFM height image of a character pair in patterned photoresist on a surface; Z scale is 700 nm. C) SEM micrograph of a character pair etched into a Si substrate. D) Section analysis of a contact-mode AFM height image of a character pair etched into a Si substrate.

A layer of AZ3312 positive photoresist (Clariant, Somerville, NJ) was spin coated, and prebaked at 90 °C for 1 min on Si substrates with a 100-nm-thick oxide layer. The wafers were then exposed through the photolithographic mask using a Karl Suss aligner and developed for 30 s in undiluted AZ300 developer (Clariant). A post-exposure bake at 110 °C for 20 min followed. The wafers were immersed in buffered HF solution for 30 s to etch the character patterns 30 nm into the silicon dioxide layer. Photoresist was stripped from the wafers using hot acetone and hot potassium hydroxide solution, and the wafers were rinsed with water. Each prepared wafer contained 81 molecular localization substrates with 9 mm x 9 mm dimensions; these pieces were separated and removed from the Si wafers using a diamond scribe. Individual localization substrates were cleaned in freshly prepared, boiling piranha solution (7:3 H_2SO_4:H_2O_2) just before use for DNA alignment. **Fig. 1** shows AFM and SEM characterization of molecular localization substrates.

Nanofabrication Follow-Up

We used these molecular localization substrates to perform a number of nanofabrication experiments to gain a deeper understanding of DNA-templated silver nanowire synthesis. Our procedures for nanofabrication have been described elsewhere,[5,9-10] and can be summarized in the following steps:

1. Initial tapping-mode AFM characterization of surface-aligned DNA.
2. Offline metallization of the DNA molecules.
3. Subsequent AFM and SEM characterization of the metallized DNA.

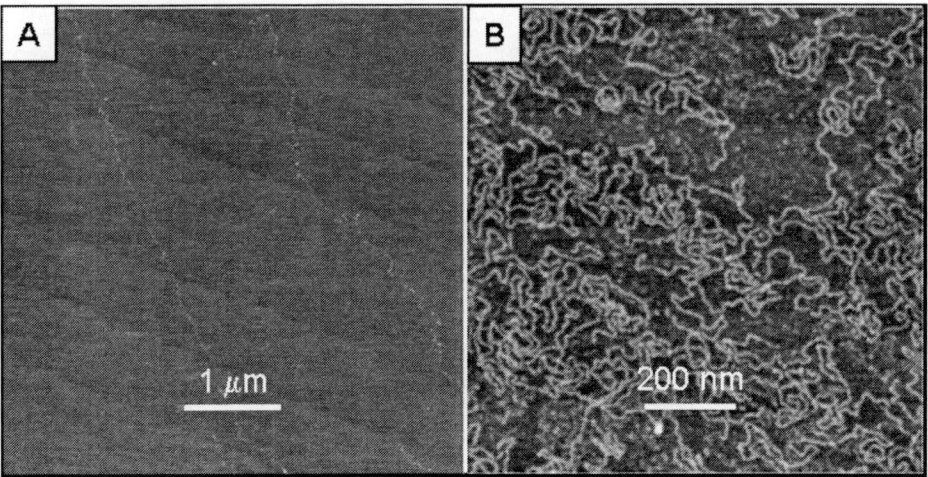

FIGURE 2. Tapping-mode AFM height images of A) surface-aligned single-stranded (ss) DNA, and B) unaligned dsDNA. Z scale is 3 nm in both cases.

DNA Alignment

Deposition and alignment of DNA molecules on the micropatterned substrates are accomplished by first treating the surface with a 1 μg/mL aqueous solution of poly-L-lysine (PLL), a positively charged polypeptide, which associates with negative charges on the silicon dioxide surface. After the surface is rinsed with water and dried under a stream of compressed air, a 0.8 μL droplet of a 1 ng/μL solution of λ DNA is translated across the substrate with controlled velocity by means of a computer-actuated two-axis translation stage. The positive charges on the PLL attract negative charges on the DNA phosphate backbone, enabling DNA immobilization on the surface. When a DNA molecule sticks to the surface at one point, the receding meniscus aligns the remainder of that molecule in the direction of droplet translation.[11-12] After DNA deposition the substrates are rinsed with water and dried under compressed air before AFM imaging. Surface-aligned DNA is attached robustly and cannot be removed even by extensive water rinsing.[9] **Fig. 2** shows typical AFM height images of aligned and non-aligned DNA molecules.

Silver metallization of dsDNA and ssDNA aligned on silicon dioxide is a common offline chemical modification carried out in our group.[10] The procedure comprises three stages: (a) self-assembly of Ag^+ along the DNA; (b) reduction of Ag^+ to create nucleation sites for further Ag deposition; and (c) developing to deposit nanoscopic amounts of Ag metal on the DNA template. The use of microfabricated molecular localization substrates for DNA alignment and metallization has enabled the study of the same molecules at multiple stages of the fabrication process, as shown in **Fig. 3**.

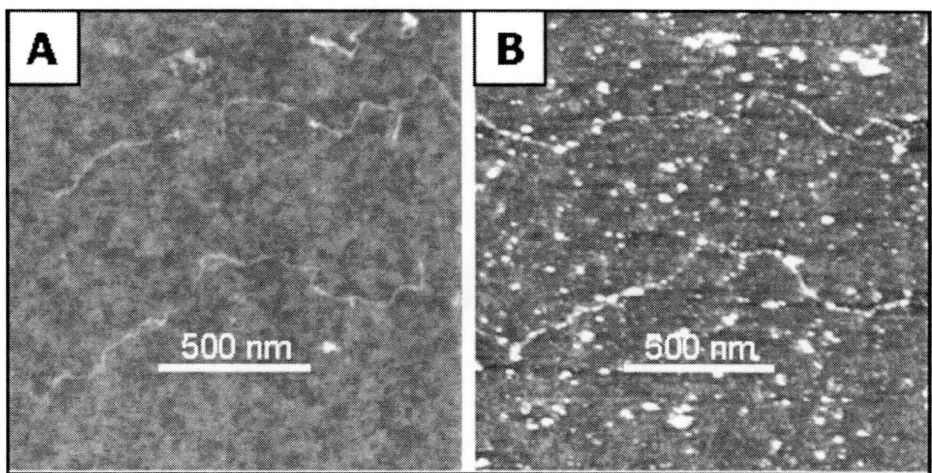

FIGURE 3. AFM characterization of a pair of ssDNA molecules on a microfabricated substrate. A) before, and B) after a silver metallization treatment. Z scale is 3 nm in both images.

In **Fig. 3** we observe that silver nanoparticles grow along the ssDNA, but nanoparticles also deposit at random locations on the surface, unassociated with the DNA. Analysis of multiple before/after images similar to **Fig. 3** has led us to hypothesize that nonspecific deposition of silver is mainly due to the presence of negative charge sites on the silicon dioxide layer, which causes aggregation of Ag^+. Upon silver reduction, these spurious self-assembly sites lower the DNA specificity of the silver deposition and the quality of the surface. Thus, the ability to study the exact same location on substrates proved instrumental in demonstrating that the nonspecific deposition of silver was not correlated with the presence of physical structures or contaminants already present on the surface. These observations led to the development of an ionic masking methodology to decrease nonspecific deposition in the fabrication of DNA-templated silver nanowires.[10]

Microfabricated molecular localization substrates can also be used in SEM characterization. In this way, a metallized DNA molecule can be re-located and analyzed to obtain information not available through AFM. For example, SEM can provide more accurate width measurements of DNA-templated nanowires than AFM (**Table 1**). Moreover, SEM offers the capability of obtaining chemical composition information through energy dispersive or wavelength dispersive X-ray analysis (**Fig. 4**).

TABLE 1. ssDNA-Templated Silver Nanowire Width Measurements by AFM and SEM.

AFM Width (nm)	SEM Width (nm)	AFM - SEM (nm)	AFM/SEM Ratio
75	56	19	1.3
57	22	35	2.6
72	29	43	2.5
38	19	19	2.0
36	17	19	2.1
111	56	55	2.0
52	29	23	1.8
65	27	38	2.4
36	22	14	1.6
44	15	29	2.9
158	102	56	1.5
46	22	24	2.1
76	34	42	2.2
35	17	18	2.1
45	17	28	2.6
32	19	13	1.7
35	12	23	2.9
54	17	37	3.2
35	15	20	2.3
93	41	52	2.3
54	27	27	2.0
57	22	35	2.6
89	49	40	1.8
121	75	46	1.6
57	29	28	2.0
63	19	44	3.2
44	19	25	2.3
22	15	7	1.5
59	24	35	2.4
33	12	21	2.7
43	19	24	2.2
57	24	33	2.3
43	15	28	3.0
33	17	16	1.9
62	22	40	2.8
Average AFM Tip-Induced Broadening		30 (additive)	2.2 (multiplicative)
Standard Deviation		12	0.49

37

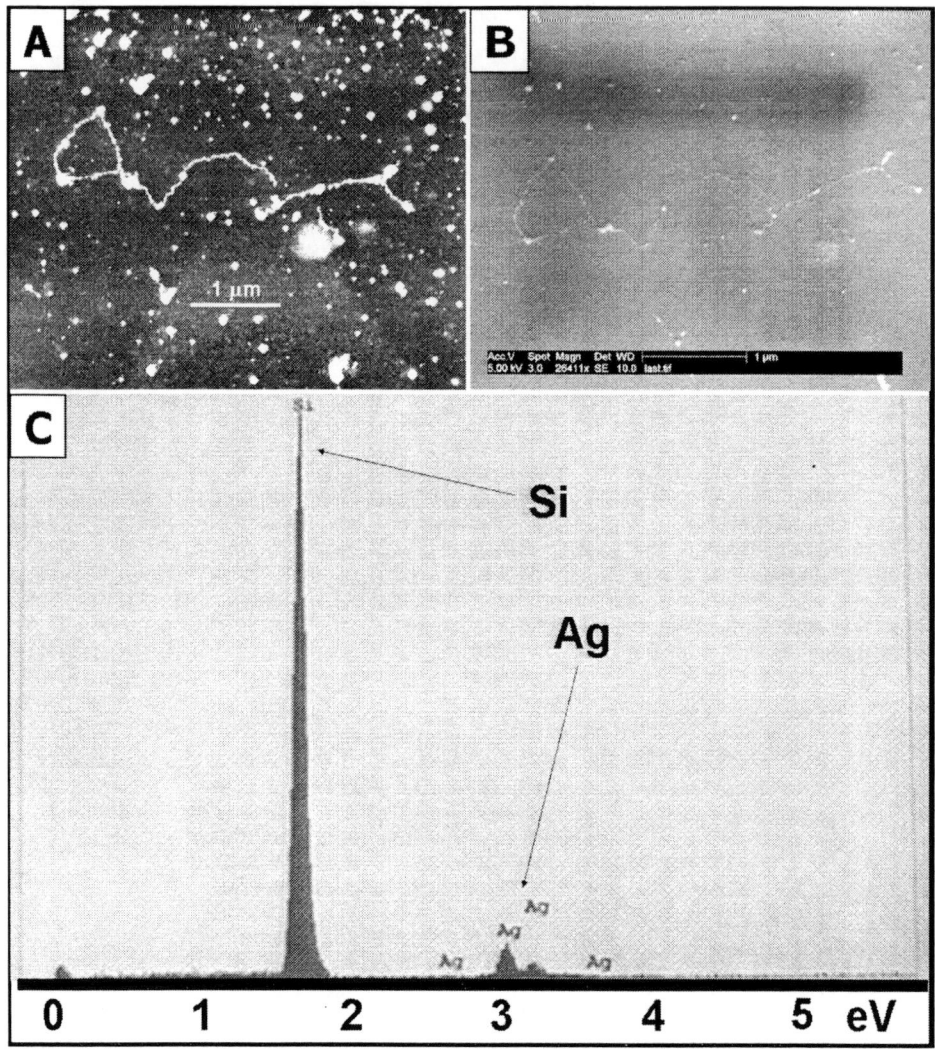

FIGURE 4. Characterization of the same metallized ssDNA molecule using A) AFM (Z scale is 10 nm) and B) SEM (5 kV acceleration potential). SEM shows that not all of the surface features from AFM have the same composition, since only grains composed of metallic silver have enough contrast to appear in SEM. C) Energy dispersive X-ray spectrum obtained from metallized DNA shows silicon (substrate) and silver (nanowire) signals.

Nanowire Annealing

Lastly, the ability to repeatedly locate the same DNA-templated nanowire on a macroscopic substrate has been utilized to conduct nanowire annealing experiments and to observe the effect of annealing on a DNA-templated nanostructure. **Fig. 5** demonstrates how increasing the annealing temperature from 200 to 500 °C modifies the grain morphology of a DNA-templated silver nanowire.

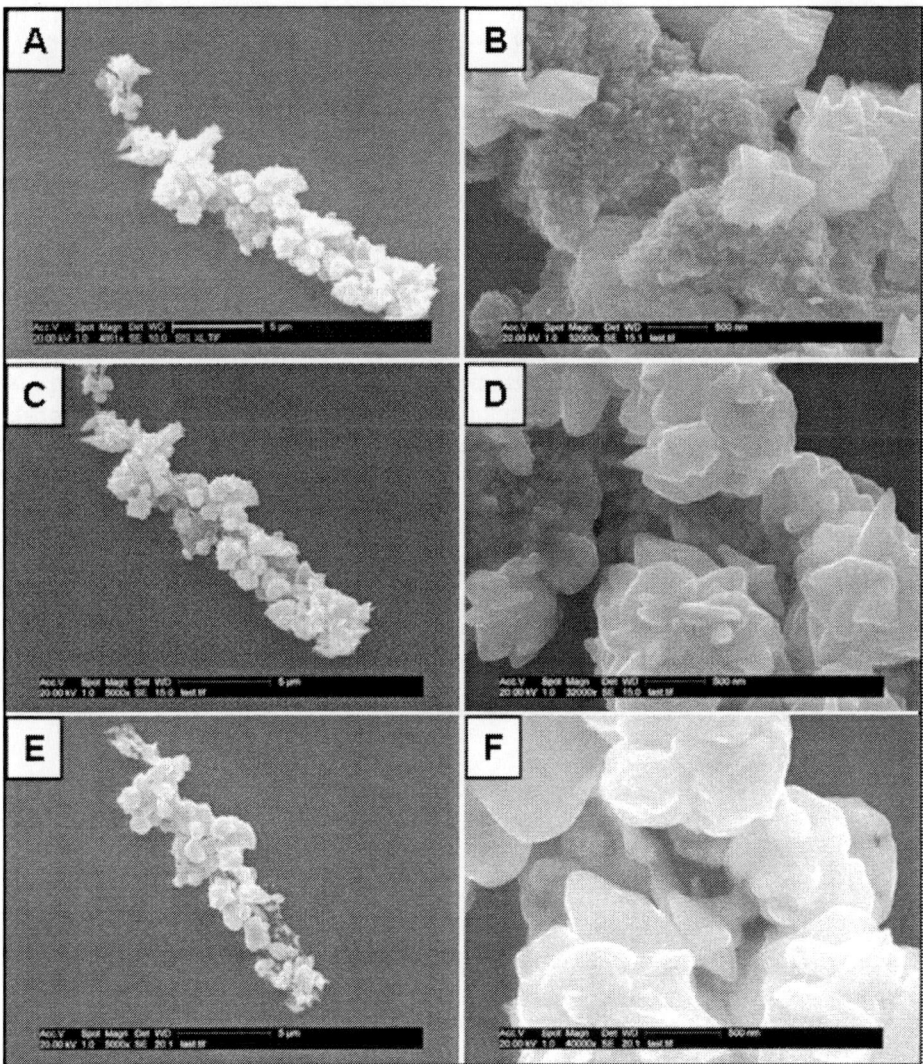

FIGURE 5. SEM characterization of the effect of thermal annealing on the structure of a λ-DNA-templated silver nanowire. A)-B) Before annealing; C)-D) after annealing 3 hrs at 200 °C; E)-F) after annealing 3 hrs at 500 °C. Thermal treatments were performed in a mixed Ar:H$_2$ (3:2 ratio) atmosphere.

CONCLUSION

A simple, rationally designed microfabricated substrate has been demonstrated for repeated single-molecule imaging by AFM and SEM. The platform is inexpensive and can be used for routine imaging since the character pair pattern does not interfere with normal sample preparation. Use of these substrates in DNA-templated nanofabrication allows the repeated characterization of single DNA molecules and other similarly sized structures at multiple stages in a nanofabrication process. The potential for investigating changes in the size, shape and composition of any individual nanostructure on the surface of the microfabricated substrate constitutes an important advantage in achieving a more fundamental understanding of the results of any nanofabrication procedure. This increased level of control over nanometer-scale construction methods should translate into better DNA-templated nanofabrication protocols and improved nanodevice quality.

ACKNOWLEDGMENTS

We wish to thank Ryan T. Kelly and Tao Pan for providing training in clean room techniques. Microfabrication work was performed in the Integrated Microelectronics Laboratory at Brigham Young University. We acknowledge partial support of this work from Brigham Young University and the donors of the Petroleum Research Fund, administered by the American Chemical Society. This work was also supported in part by the Army Research Laboratory and the U.S. Army Research Office under grant number DAAD19-02-1-0353.

REFERENCES

1. Drexler, E., *Unbounding the Future*. New York: Quill (1993).
2. Shivashankar, G. V. and Libchaber, A., *Appl. Phys. Lett.* **71**, 3727–3729 (1997).
3. Samorì, B., *Angew. Chem. Int. Ed.* **37**, 2198–2200 (1998).
4. Hansma, H. G., *Annu. Rev. Phys. Chem.* **52**, 71–92 (2001).
5. Monson, C. F. and Woolley, A. T., *Nano Lett.* **3**, 359–363 (2003).
6. Scheibel, T., Parthasarathy, R., Sawicki, G., Lin, X. -M., Jaeger, H. and Lindquist, S. L., *Proc. Natl. Acad. Sci., USA* **100**, 4527–4532 (2003).
7. Farneth, W. E., McLean, R. S., Bolt, J. D., Dokou, E. and Barteau, M. A., *Langmuir* **15**, 8569–8573 (1999).
8. Mulvaney, P. and Giersig, M., *J. Chem. Soc. Faraday Trans.* **92**, 3137–3143 (1996).
9. Woolley, A. T. and Kelly, R. T., *Nano Lett.* **1**, 345–348 (2001).
10. Becerril, H. A., Stoltenberg, R. M., Monson, C. F. and Woolley, A. T., *J. Mater. Chem.* **14**, 611–616 (2004).
11. Li, J., Bai, C., Wang, C., Zhu, C., Lin, Z., Li, Q. and Cao, E., *Nucleic Acids Res.* **26**, 4785–4786 (1998).
12. Deng, Z. and Mao, C., *Nano Lett.* **3**, 1545–1548 (2003).

DNA SUPERSTRUCTURES

Self-assembled DNA Structures for Nanoconstruction

Hao Yan, Peng Yin, Sung Ha Park, Hanying Li, Liping Feng,
Xiaoju Guan, Dage Liu, John H. Reif, Thomas H. LaBean

Department of Computer Science, Duke University, USA

Abstract. In recent years, a number of research groups have begun developing nanofabrication methods based on DNA self-assembly. Here we review our recent experimental progress to utilize novel DNA nanostructures for self-assembly as well as for templates in the fabrication of functional nano-patterned materials. We have prototyped a new DNA nanostructure known as a cross structure. This nanostructure has a 4-fold symmetry which promotes its self-assembly into tetragonal 2D lattices. We have utilized the tetragonal 2D lattices as templates for highly conductive metallic nanowires and periodic 2D protein nano-arrays. We have constructed and characterized a DNA nanotube, a new self-assembling superstructure composed of DNA tiles. We have also demonstrated an aperiodic DNA lattice composed of DNA tiles assembled around a long scaffold strand; the system translates information encoded in the scaffold strand into a specific and reprogrammable barcode pattern. We have achieved metallic nanoparticle linear arrays templated on self-assembled 1D DNA arrays. We have designed and demonstrated a 2-state DNA lattice, which displays expand/contract motion switched by DNA nanoactuators. We have also achieved an autonomous DNA motor executing unidirectional motion along a linear DNA track.

INTRODUCTION

DNA is an extraordinarily versatile material for designing nano-architectural motifs, due in large part to its programmable G-C and A-T base pairing into well-defined secondary structures. These encoded structures are complemented by a sophisticated array of tools developed for DNA biotechnology: DNA can be manipulated using commercially available enzymes for site-selective DNA cleavage, ligation, labeling, transcription, replication, kination, and methylation. DNA nanotechnology is further empowered by well-established methods for purification and structural characterization and by solid-phase synthesis, so that any designer DNA strand can be constructed. The above advantages of DNA as a nanoconstruction material explain the rapid and exciting progress in DNA based nanotechnology in recent years [1-3], especially in self-assembled nanostructures [4-6], nanorobotics [7-16], and nanocomputation [17-24]. In this paper, we review our recent experimental progress in constructing novel self-assembled DNA structures for nanofabrication and nanorobotics, and we also discuss their applications in nanotechnology, with an emphasis on DNA-based nannoelectronics.

CP725, *DNA-Based Molecular Electronics: International Symposium,* edited by W. Fritzsche
© 2004 American Institute of Physics 0-7354-0206-X/04/$22.00

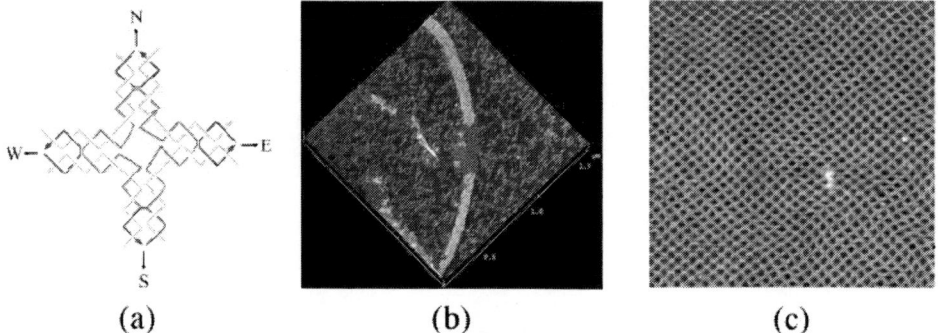

(a)   (b)   (c)

FIGURE 1 (a) Strand structure of the cross DNA nanostructure (4x4 DNA tile). **(b)** Atomic Force Microscope (AFM) image of a nanoribbon self-assembled from the cross DNA nanostructure tiles. **(c)** AFM image of a nanoribbon composed of 4x4 tiles. Scale: 1.5 um x 1.5 um **(e)** AFM image of a 2D nanogrid self-assembled from the cross structure tiles. Scale: 600 nm x 600 nm.

SELF-ASSEMBLED DNA LATTICES

DNA Nanogrids and Nanoribbons

Self-assembling nanostructures composed of DNA molecules offer great potential for bottom-up nanofabrication of materials and objects with nanometer scale features. Potential applications of DNA self-assembly and scaffolding include nanoelectronics, biosensors, and programmable/autonomous molecular machines. We have recently achieved the design, characterization and self-assembly of a novel DNA nanostructure (referred to as DNA cross structures or 4x4 tiles) [25]. This planar tile consists of four four-armed branch junctions pointing at four directions (north, east, south, and west in the tile plane). Special features of this tile structure include a square aspect ratio, which may help regularize lattice growth by balancing helix stacking and sticky-end connections in all four directions within the lattice plane. Figure 1a gives a schematic drawing of the strand structure of the 4x4 tile. Note that a central strand weaves through all the four four-armed junctions. Though each arm is of flexible Holliday junction structure, when combined with junctions on neighboring tiles they are able to form reasonably rigid nanostructures.

With slight modification of the tile spacing and its sticky-end association, we are able to program the self-assembly of the above 4x4 tiles into two distinct lattice morphologies: a uniform width nanoribbon and a 'waffle-like 2D planar nanogrid structure. A nanoribbon is formed when the identical face of every constituent tile points toward the same direction in the lattice; the planar grid, in contrast, is formed when the identical face of each adjacent tile points up and down alternatively. Figures 1b and c give AFM images of the 1D nanoribbon and 2D nanogrid.

The DNA nanogrid and nanoribbon can serve as scaffolds to organize other molecular components. In particular, we have achieved a) periodic streptavidin nanoarrays templated on 2D nanogrids via affinity binding to biotin labeled DNA strands, and b) highly conductive sliver and gold nanowires via electroless deposition. Details can be found in ref. 25.

DNA Lattice Nanotubes from TX Tiles

We have recently reported results on the construction and characterization of DNA nanotubes, a new self-assembling superstructure composed of DNA tiles. Triple-crossover (TX) tiles modified with thiol-containing dsDNA stems projected out of the tile plane were utilized as the basic building block. TX nanotubes display a constant diameter of approximately 25 nm and have been observed with lengths up to 20 microns.

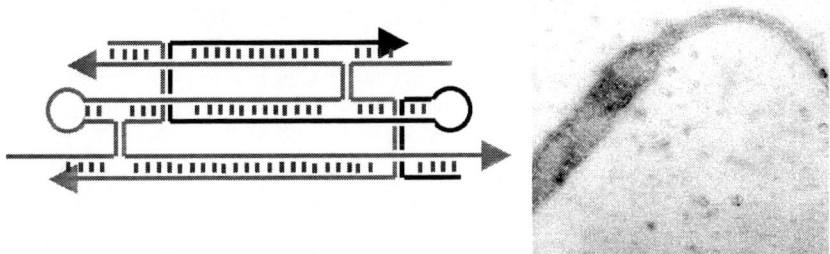

FIGURE 2. *Left panel:* Schematic drawing of a TAO DNA tile. *Right panel:* TEM image of a DNA nanotube constructed from two different tile types.

Figure 2 shows a negative stained transmission electron microscopy (TEM) image of a section of TX DNA nanotube with the left side apparently unwrapped while the right side remains in a tight tube-like structure. Lighter colored bands are visible and identified as protruding stem-loops on the B tiles of the 2-tile AB set. The bands have spacing of approximately 28 nm, in good agreement with the design. Other high resolution images of the constructs from TEM and AFM as well as preliminary data on successful metallization of the nanotubes have been published [26]. DNA nanotubes represent a potential breakthrough in the self-assembly of nanometer scale circuits for electronics layout since they can be targeted to connect at specific locations on larger-scale structures and can subsequently be metallized to form nanometer scale wires. The dimensions of these nanotubes are also perfectly suited for applications involving interconnection of molecular scale devices with macroscale components fabricated by conventional photolithographic methods.

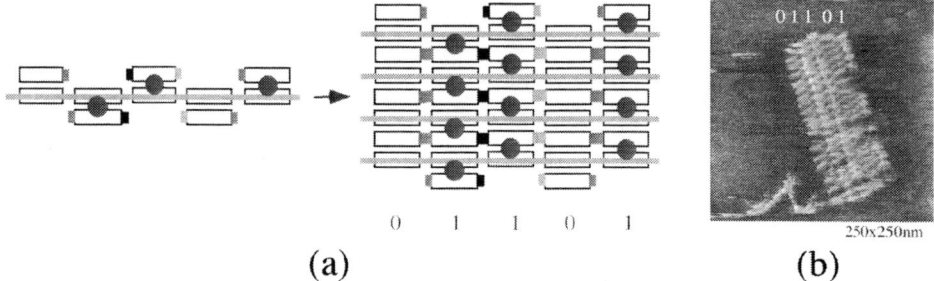

(a)   (b)

FIGURE 3 (a) Simplified schematic drawing of the self-assembly of the DNA barcode lattice 01101. The gray thick line represents a scaffold strand weaving continuously through five separate DNA tile structures. Each black dot represents a DNA stem loop coming out from its corresponding DNA tile. Multiple layers of such structure will associate with each other via sticky end pairings and form a 2D grid displaying the barcode information, which can be detected by AFM. **(b)** AFM image of a lattice displaying the barcode information of 01101. 1 and 0 bit values are clearly visible as lighter and darker stripes.

Directed Nucleation of Barcode DNA Lattices

We recently reported the construction of an aperiodic patterned DNA lattice (barcode lattice) formed by a self-assembly process via the directed nucleation of DNA double crossover (DX) tiles [28] around a scaffold DNA strand [27]. Figure 3a shows a schematic drawing of the self-assembly of a barcode lattice representing bit values of 01101. A scaffold strand (shown as gray thick line in Figure 2a) encodes information 01101, and serves as the nucleation point for the assembly of DX tiles, with each bit represented by a DX tile. To aid in visual read-out of the encoded information 01101, each bit 1 is displayed as two stem loops perpendicular to the tile plane, with one protruding upward and the other downward; each bit 0, in contrast, is represented as the absence of such stem loops. Multiple layers of such structure will associate with each other via sticky ends pairing and form a 2D grid displaying the barcode information, which can be easily detected by AFM. Figure 3b shows one such image.

We have also demonstrated the reprogramming of the system to another patterning; an inverted barcode pattern of 10010 was achieved by modifying the scaffold strands and one of the strands composing each DX tile. A ribbon lattice, consisting of repetitions of the barcode pattern with expected periodicity, was also constructed by the addition of sticky-ends. The patterning of both classes of lattices was clearly observable via atomic force microscopy. These results represent a first step toward implementation of a visual readout system capable of converting information encoded on a 1D DNA strand into a 2D form readable by advanced microscopic techniques. A functioning visual output method would not only increase the readout speed of DNA-based computers, but may also find use in other sequence identification techniques such as mutation or allele mapping. Details of the barcode lattice construction and analysis can be found in ref. 27.

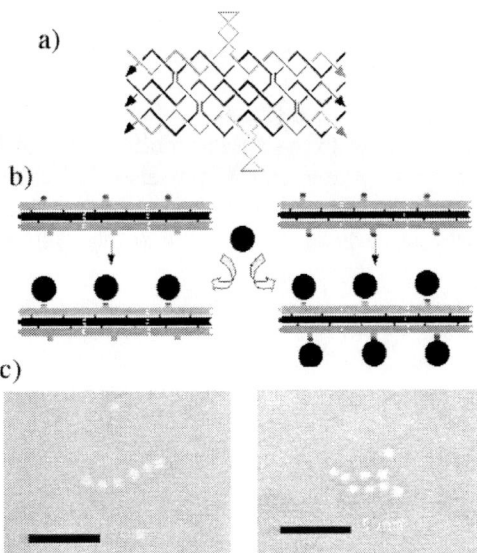

FIGURE 4 (a) Strand structures of triple crossover DNA tile with two stem loops protruding out the tile on two sides. **(b)** schematic drawings of the TX DNA-templated self-assembly of single-layer and double-layer streptavidin-gold linear arrays. Left: streptavidin-gold binds to one side of the 1D TX arraysingle layer through streptavidin-biotin interatction to form single layer gold nanoparticle arrays; Right: streptavidin-gold binds to both side of the 1D TX arraysingle layer through streptavidin-biotin interatction to form double layer gold nanoparticle arrays. Streptavidin-gold is represented by black ball. **(c)** scanning electron microscopy (SEM) images of single-layer (left panel) and double-layer (right panel) streptavidin-gold arrays. Scale bars: 50 nm.

DNA Templated Linear Nano-Particle Arrays

Self-assembled DNA arrays provide an excellent template for spatially positioning other molecules with increased relative precision and programmability. One potential application of DNA nanotechnology is the use of self-assembled DNA lattices to scaffold assembly of nanoelectronic components, especially metallic nanoparticles. We have recently demonstrated the use of a linear triple crossover (TX) DNA array for the assembly of streptavidin conjugated 5 nm gold nanoparticles, where the gold can be precisely positioned periodically on the self-assembled DNA array [29]. Two forms (single-layer and double-layer) of streptavidin-coated gold nanoparticle linear arrays were achieved on DNA template. In this system, each TX tile (Figure 4a) is designed to contain two stem loops protruding on two opposite sides of the TX molecule in the tile plane. To obtain a single-layer streptavidin-gold array, only the stem loop on one side of each TX tile is modified with biotin; to obtain double-layer of strepatividin-gold array, stem loops on both sides of each TX tile are modified with biotin. Streptavidin coated gold nanoparticles (shown as dark black balls in Figure 4b) specifically bind to the biotin-modified stem loops, and thus self-organize into single- or double-layer arrays respectively. Figure 4c shows scanning electron microscopy (SEM) images demonstrating the TX array templated self-assembly of single layer and double layer streptavidin-gold particles. The distance between each pair of adjacent gold particles in the single layer and double layer arrays is ~17 nm, matching the designed parameter.

The use of DNA nanostructure to organize nanoparticles into programmable arrays could provide a powerful tool to assemble architectures for nanoelectronic device construction and electrical measurements. It may also aid in building logical molecular electronic devices such as quantum cellular automata. In addition, an array of uniformly gapped metallic particles might serve to interconnect other nanoelectronic components.

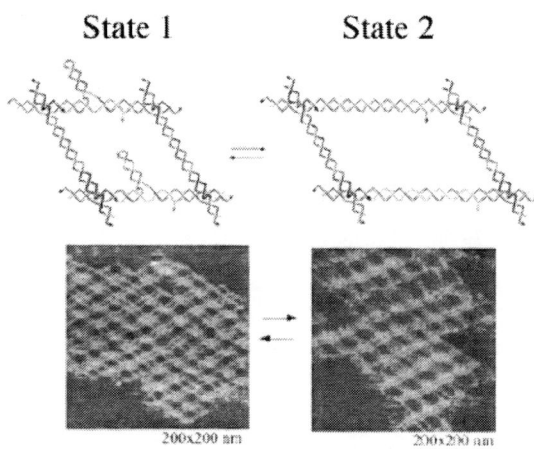

FIGURE 5 *Upper panel:* Schematic drawing of the two states of the lattice unit structure. *Lower panel:* Representative AFM images of the rhombus lattices displaying different morphologies, which can be reversibly switched by the DNA nanoactuator.

DNA NANOMECHANICAL DEVICES

A Two-state DNA Lattice Switched by DNA Nanoactuator

Controlled mechanical movement in molecular-scale devices is one of the key goals of nanotechnology. DNA is an excellent candidate for the construction of such devices due to the specificity of base pairing and its robust physicochemical properties. A variety of DNA-based molecular machines displaying rotational and open/close movements have recently been demonstrated [reviewed in ref. 30]. Reversible shifting of equilibrium between two conformational states is triggered by changes in experimental conditions or by the addition of a " DNA fuel strand" that provides the driving force for such changes. Incorporation of DNA devices into arrays could lead to complex structural states suitable for nanorobotic applications, if each individual device can be addressed separately.

We have recently reported the construction of a robust sequence dependent DNA device, which we call a "nano-actuator" and the incorporation of such devices into a 2D parallelogram DNA lattice [31]. The nanoactuator can exist in two states. State 1 is the shortened state with a bulged three-arm DNA branch junction; state 2 is the elongated state with two perfectly complementary strands of DNA. Bulged 3-arm DNA branch-junctions have been well characterized and extensively used in DNA nanoconstruction and as topographic markers in self-assembly of 2D DNA lattices. Thus, a DNA device based on bulged three-arm junction is an excellent candidate to serve as actuator for DNA lattices. The parallelogram lattice contained one such device two opposite edges of each unit cell (Figure 5). Large alterations in lattice dimension were observed due to the additive changes from each unit cell. Lower panel of Figure 5 shows the AFM images demonstrating the inter-conversion of the two states of the rhombus lattice actuated by the actuator devices. The sizes of the cavities in the rhombus lattice were switched from ~ 14 nm x 14 nm (the left image) to ~ 14 nm x 20 nm (the right image). Reverse process from extended lattice to contracted lattice was also observed. Details of the 2-state lattice system are reported in ref. 31.

The above switchable DNA lattices promise numerous potential applications in nanofabrication, nanocomputation, and nanoelectronics. One particularly attractive application could be the controlled nanofabrication of molecular nanoelectronic wires with 'on' and 'off' states, since the size and the shape of the underlying DNA lattice templates can be programmably controlled using DNA sequence dependent nanoactuator devices.

An Autonomous DNA Walker Moving Along a Track

Most molecular machines executing cellular functions in human body are autonomous and in many cases unidirectional, which makes the construction of such autonomous unidirectional devices in artificial systems promising and attractive. We have recently reported the design and experimental construction of an autonomous unidirectional DNA walker along a DNA track [32]. This walker device has the following features: 1) The device is free of any external environmental mediations, and hence is autonomous. It is powered by the hydrolysis of ATP consumed by T4 ligase. 2) The motion of the device is unidirectional. 3) The device executes motion along a DNA track and renders a DNA fragment moving unidirectionally from one end of the track to the other end.

The structural design and the operation of the device are shown in Figures 6. The walker device is composed of two parts: the 'track' and the 'walker'. The track consists of three evenly spaced DNA double helical 'anchorages' (A, B, and C), each tethered to another DNA double helical segment 'backbone' via a flexible single strand DNA 'hinge'; the walker is a 6-nucleotide DNA initially residing on top of anchorage A (labeled * and represented as bold fragments in Figure 6).

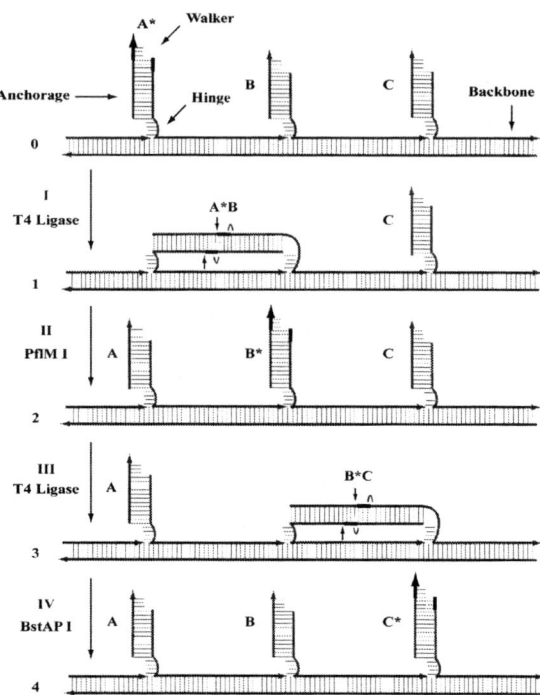

FIGURE 6 The structural design and operation of the autonomous unidirectional device. The bold fragments are the walker fragments. The arrows indicate the PflM I and BstAP I restriction sites. The curves indicate the ligation sites.

The walker moves sequentially along the track from anchorage A to B, then to C in an autonomous, unidirectional fashion. The motion is facilitated by the alternate actions of ligase and restriction endonucleases, and powered by the hydrolysis of ATP. Panel 0 of Figure 6 shows the walker (indicated by *) residing at anchorage A before the motion starts. At this point, the anchorage/walker complex A* and the anchorage B have complementary sticky ends, which will hybridize with each other. The nicks in the hybridization complex A*B are subsequently sealed by T4 ligase, which covalently joins the two anchorages (process I in Figure 6). This is an irreversible step driven by the energy released from the hydrolysis of ATP. Next, a restriction endonuclease PflM I recognizes the newly generated recognition site in A*B and cuts the walker to anchorage B* (process II). Now the newly generated sticky end of B* is complementary to that of C, and the two will hybridize with each other. Again, T4 ligase seals the nicks (process III), producing a new recognition site for another endonuclease BstAP I, which will cut the walker to anchorage C, completing the autonomous unidirectional movement of the walker (process IV). The above autonomous, unidirectional operation of the walker device was verified via the careful tracking of the radioactively labeled walker using gel electrophoresis. For details, see ref. 32.

In principle, we can extend the motion of the walker well beyond the 3-anchorage system demonstrated above. By encoding information into the walker and the anchorages, we have accomplished a theoretical design of an autonomous nano-mechanical device capable of universal computation and hence universal translational motion [33]. It is also conceivable to embed the walking device into well defined DNA lattices [4-6, 27] and thus obtain ('intelligent') robotics lattices. Nanorobotics systems of this kind would have many applications in nano-computing, nano-fabrication, and nano-electronics.

CONCLUSION

In this review, DNA as a designer molecule for constructing self-assembled DNA lattices and nanomechanical devices was discussed. The experimental demonstrations described here further attest to DNA's role as a leading material for nanoconstruction. The fascinating potential applications of self-assembled DNA structures for nanoelectronic, nanorobotics, and nanocomputation are awaiting us to explore.

ACKNOWLEDGMENTS

The authors would like to thank National Science Foundation for kind support.

REFERENCES

1. Seeman, N.C., *Nature* **421**, 427-431 (2003).
2. LaBean, T.H., 'Introduction to self-assembling DNA nanostructures for computation and nanofabrication" in *Computational Biology and Genome Informatics*, edited by Wang, J.T.L., Wu, C.H. and Wang, P.P. World Scientific Publishing, River Edge, NJ, 2003.
3. Reif, J.H. DNA lattices: a programmable method for molecular scale patterning and computation, *Computer and Scientific Engineering Magazine*, special issue on Bio-Computation, IEEE Computer Society, 32-41 (2002).
4. Winfree, E., Liu, F., Wenzler, L.A., and Seeman, N.C., *Nature* **394**, 539-544 (1998).
5. LaBean, T.H., *J. Am. Chem. Soc.* **122**, 1848-1860 (2000).
6. Mao, C., Sun, W., and Seeman, N.C., *J. Am. Chem. Soc.* **121**, 5437-5442 (1999).
7. Mao, C., Sun, W., Shen, Z., and Seeman, N.C., *Nature* **397**, 144-146 (1999).
8. Yurke, B, Turberfield, A.J., Mills, A.P., Jr, Simmel, F.C., and Jennifer L.N., *Nature* **406**, 605-608 (2000).
9. Yan, H., Zhang, X., Shen, Z., and Seeman, N.C., *Nature* **415**,62-65, (2002).
10. Li, J.J, and Tan, W., *Nano Lett.* **2**, 315-318 (2002).
11. Alberti, P. and Mergny, J.L., *Proc. Natl. Acad Sci. USA*, **100**, 1569–1573 (2003).
12. Chen, Y., Wang, M., and Mao, C., *Angew. Int. Chem. Ed.*, In press (2004).
13. Sherman, W.B. and Seeman, N.C., *Nano. Lett.*, , (2004).
14. Simmel, F.C. and Yurke, B., *Physical Review E* **63**, 041913, (2001).
15. Simmel, F.C. and Yurke, B., *Applied Physics Letters*, **80**, 883–885, (2002).
16. Turberfield, A.J. et al.*Phys. Rev. Lett.*, **90**, 118102, (2003).
17. Adleman, L.M., *Science* **266**, 1021-1024 (1994).
18. Liu, Q. *et al, Nature* **403**, 175-179 (2000).
19. Mao, C., LaBean, T.H., Reif, J.H., and Seeman, N.C., *Nature* **407**, 493-496 (2000).
20. Benenson, Y. *et al.*, *Nature* **414**, 430 – 434 (2001).
21. Benenson, Y. *et al.*, *Nature* **429**, 423-429 (2004).
22. Benenson, Y. *et al.*, *Proc. Natl. Acad. Sci. USA*, **100**, 2191–2196 (2003).
23. Ravinderjit, B.S. *et al.*, *Science* **296**, 499-502 (2002).
24. Mirkin, C.A., *Inorg. Chem* **39**, 2258 (2000).
25. Yan, H., Park, S.H., Finkelstein, G., Reif, J.H., and LaBean T.H., *Science* **301**, 1882-1884 (2003).
26. D. Liu, S- H. Park, J. H. Reif, and T.H. LaBean *Proc. Nat. Acad. Sci., USA* **101**, 717-722 (2004).
27. Yan, H., LaBean, T.H., Feng, L., and Reif, J.H., *Proc. Natl. Acad. Sci. USA* **100**, 8103-8108 (2003).

28. Seeman, N.C. *J. Biomol. Struct. Dyn.* **8,** 573–581 (1990).
29. Li, H., Park, S.H., Reif, J.H, LaBean, T.H., and Yan, H., *J. Am. Chem. Soc.* **126,** 418-419 (2004).
30. Niemeyer, C. and Adler, M., *Angew. Chem.-Int. Edit.* **41,** 3779-3783 (2002).
31. Feng, L., Park, S.H., Reif, J.H., and Yan, H., *Angew. Chem.-Int. Edit.* **42,** 4342-4346 (2003).
32. Yin, P., Yan, H., Daniell, X.G., Turberfield, A.J., and Reif, J.H., *Angew. Chem.-Int. Edit.* In press (2004).
33. Yin, P., Turberfield, A.J., Sahu, S.,and Reif, J. H., 10th International Meeting of DNA Based Computers (2004).

Coupling G-Wires To Metal Nanoparticles

Claudia Holste, Anett Sondermann, Robert Möller, Wolfgang Fritzsche

Institute for Physical High Technology, POB 100239, 07702 Jena, Germany

Abstract. G-wires are DNA superstructures based on quartet formation by four guanine (G) bases. These molecular structures reach the micrometer size scale by assembling short oligonucleotides of a guanine rich sequence. Physico-chemical parameters were optimized to achieve optimal superstructure assembly. The combination of these superstructures with metal nanoparticles could provide new tools for a molecular nanotechnology. An assembly strategy has been developed and first experiments towards an inclusion of nanoparticles in the supramolecular assembly were conducted.

INTRODUCTION

DNA Nanotechnology

DNA nanotechnology can be divided into two parts based on the length of the applied molecules: The oligonucleotide-based part includes DNA chip technology and the bioconjugation of inorganic nanoparticles, and the other part is dealing with rather long (usually some micrometers) molecules. The former is based on larger molecule numbers, the latter often working with individual molecules. Regarding the size, oligonucleotides are hardly visible for ultramicroscopic techniques, but long DNA can be routinely imaged in an extended state. DNA nanotechnology promises to solve the integration problem in molecular nanotechnology, and this potential is based on the combination of both worlds: Long DNA provides the anchor to the macroscopic world by connection to microelectrodes etc, and short DNA offers the specific „glue" to include other DNA, biomolecules or even inorganic materials.

DNA superstructures to bridge nano- and microworld

However, long DNA is often difficult to work with, and the positional control of the immobilization rather difficult. As mentioned before, short DNA is not easily accessible by single molecule characterization techniques. A solution to this dilemma provide DNA superstructures based on specific DNA-DNA interactions of short DNA units. One example is the extended work of Seeman et al., using the concept of reciprocal exchange between DNA double helices or hairpins to produce branched DNA motifs, like Holliday junctions, or related structures, such as double crossover, triple crossover, paranemic crossover and DNA parallelogram motifs [1]. So micrometer structures are generated out of molecular units, and can be applied for the

CP725, DNA-Based Molecular Electronics: International Symposium, edited by W. Fritzsche
© 2004 American Institute of Physics 0-7354-0206-X/04/$22.00

integration of other materials such as nanoparticles and proteins [2]. Another approach is based on G-quartets, a structure formed by four Guanine bases situated in a plane and connected by hydrogen bonds. Such quartets (intramolecular) are known for the G-rich chromosome ends (telomeres), where they help to protect single-stranded overlaps against degradation. Using G-rich short oligonucleotides, DNA superstructures have been assembled based on this interaction, named G-wires [3]. These G-wires are an interesting material for nanotechnology due to their underlying DNA backbone and their increased stiffness and proposed improved electrical conductivity compared to double-stranded DNA

MATERIALS & METHODS

G-Wire

DNA oligonucleotides with the sequence GGGGTTGGGG were incubated in a 125 ng/µl concentration in 50 mM NaCl, 10 mM MgCl$_2$ and 50 mM Tris-HCl (pH 7.5) [4]. For testing the effect of ions, this buffer contained 25 mM NaCl and 25 mM Kcl beside MgCl$_2$ and Tris-HCl. The assembled wires were diluted (10-100 fold) in 1 mM MgCl$_2$, 10 mM Tris-HCl (pH 7.5) and 3 µl droplets incubated on a mica substrate for 5 min. After washing with aqua dest, the samples were dried in a stream of N$_2$ gas.
For the conjugation of streptavidin-modified nanoparticles, DNA with a terminal biotin was added.

Gold Nanoparticles

Gold nanoparticles of 5 nm diameter conjugated to streptavidin (British Biocell) were used for the experiments. After G-wire immobilization, the samples were incubated with gold nanoparticles prior to extensive washing.

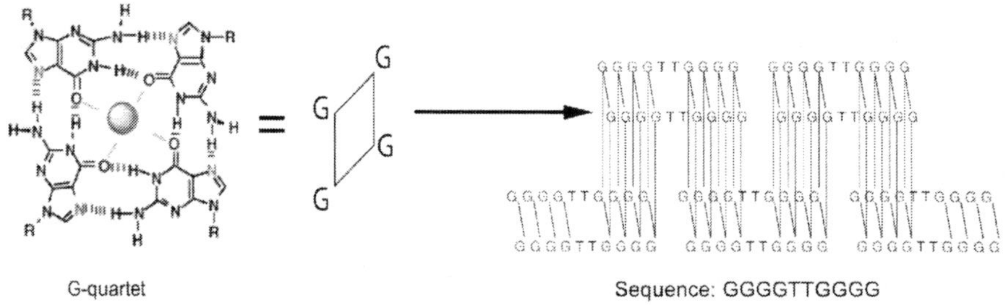

G-quartet

Sequence: GGGGTTGGGG

FIGURE 1. Principle of G-wire formation. The formation of a planar arrangement of 4 Guanine bases connected by hydrogen bonds (left) can be observed and is denoted as G quartet. Such G quartets can be generated intramolecular by backfolding or (as shown on the right) between G-rich DNA molecules. In the case of short oligonucleotides a superstructure called G-wires is formed.

Scanning Force Microscopy

Scanning force microscopy was with either a Multimode or a Dimension 3100 measurement head of a NanoScope III (Digital Instruments, Santa Barbara, CA) in tapping mode in air.

RESULTS & DISCUSSIONS

Effect Of Ions Onto G-Wire Formation

The assembly of G-wires requires the presence of Guanin-rich oligonucleotides and ions. Typically, mono- and divalent cations play an essential role in both the formation and the subsequent immobilization, because these processes rely on electrostatic interactions. A first line of experiments (Fig. 2) addressed the composition of the applied buffers regarding the mentioned ions, omitting NaCl (a), NaCl/KCl (b), Tris-HCl (c) or $MgCl_2$ (d), respectively. Best results were achieved without NaCl/KCl, in this case extended G-wires became visible as thread-like structures in the topographic contrast of the AFM (Fig. 1b). The absence of either NaCl (a) or Tris-HCl (c) leads to a reduced growth, but based on height and shape the observed structures still point to short G-wires. Only in the case of the experiment without $MgCl_2$, the imaged substrate is free of G-wires (d). However, because divalent 'bridging' ions like Mg^{2+} are thought to be essential for attaching DNA on mica, this result can be explained by failing adsorption of the G-wires.

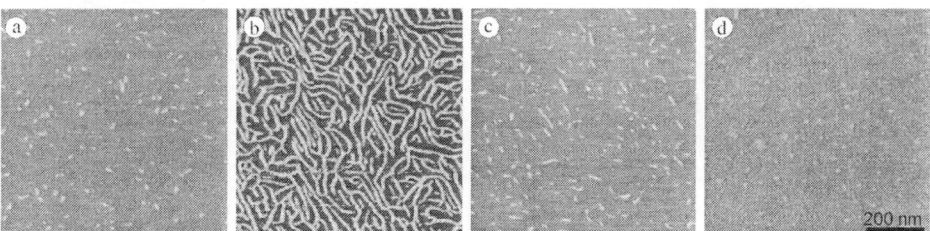

FIGURE 2. Influence of ions on the formation of G-wires. The following salts were omitted in the standard buffer: a) NaCl, b) NaCl/KCl, c) Tris-HCl, d) $MgCl_2$.

Concentration of Oligonucleotides

In order to establish optimal experimental conditions, the starting concentration for the G-containing oligonucleotides had to be determined. G-wire growth is a dynamic process, and so one would expect a minimal DNA concentration in the solution to get a reasonable yield of superstructures. The surface density of G-wires in the AFM is influenced by both the solution concentration and the adsorption time for a given immobilization protocol. For a fixed set of adsorption conditions (including time), various DNA concentrations for G-wire growth were tested. Previous experiments and literature values pointed to concentrations in the µM range, and the experiments shown in Fig. 3 were conducted with concentrations of 0 (a), 62.5 (b), 125 (c) and 187,5 ng/µl (d). The features visible in Fig.3 b and c can be attributed to the added DNA, and beside more globular structures also extended assemblies are visible, especially for the higher concentration. Increasing the concentration even higher to about 185 ng/µl yields predominantly extended assemblies, and also the average length of the G-wires is clearly enhanced (d). An alignment of the structures is observed: Apparently they orient themselves at three 60° angles with the underlying mica crystal arrangement [5].

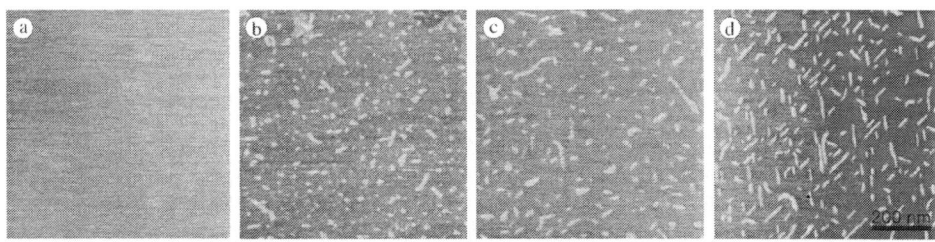

FIGURE 3. Influence of DNA oligonucleotide concentration onto G-wire assembly. Different concentrations of DNA were incubated with 50 mM NaCl, 10 mM MgCl₂ and 50 mM Tris-HCl (pH 7.5) prior to immobilization onto mica in 1 mM MgCl2 and 10 mM Tris-HCl (pH 7.5). Oligonucleotide concentration was a) 0, b) 62.5, c) 125, d) 187.5 ng/µl, respectively.

Biotin-Streptavidin System For Conjugation Of Nanoparticles To G-Wires

After establishment of optimal conditions for G-wire growth and imaging, the next step was to investigate possibilities for the inclusion of other biological or inorganic parts into the G-wires. In order to conjugate structures to biomolecules, a well-established coupling system is provided by biotin-streptavidin. So we could use biotinilated DNA molecules and try to attach streptavidin-conjugated particles by highly specific interactions. Incorporation of biotinilated DNA in G-wires was difficult, because G-wire formation is apparently hindered by the biotin modification. The first series of experiments addressed the optimal fraction of biotinilated DNA. A concentration below 5 ng/µl (from an overall DNA concentration of 125 ng/µl) was found as a compromise. Using this protocol, G-wires with included biotinilated DNA were assembled and immobilized onto mica (Fig. 4, left). Incubation with streptavidin-functionalized gold nanoparticles (diameter 5 nm) prior to washing resulted in binding of the nanoparticles onto the surface, often at the ends of G-wires (Fig. 4, right). Hardly any free G-wire was detected, pointing to an efficient binding of the conjugated gold to the DNA superstructures. However, free gold particles not connected to G-wires were also detected. This observation pointed to a surplus in nanoparticles and a substrate with some affinity to gold nanoparticles. Future experiments will adjust the G-wire / nanoparticle ratio and will address the unspecific binding of gold by the development of efficient blocking protocols.

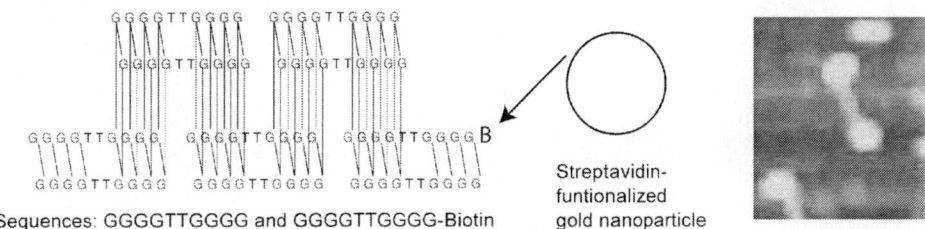

Sequences: GGGGTTGGGG and GGGGTTGGGG-Biotin

Streptavidin-
funtionalized
gold nanoparticle

FIGURE 4. Specific binding of gold nanoparticles to G-wires using biotinilated DNA and streptavidin-conjugated gold. left: Scheme of G-wires consisting of non-modified and biotinilated DNA (the "B" stands for the biotin modification). These G-wires are the framework for binding of streptavidin-conjugated colloidal gold (center). right: AFM image of G-wire structure with two gold nanoparticles bound on both ends.

Conclusions

We could demonstrate the controlled combination of DNA superstructures with inorganic and highly defined nanostructures (gold particles) as a prerequisite for further developments in the field of molecular nanotechnology. Conditions for an optimal generation of G-wires were established, biotinilated DNA as ,handle' for gold attachment introduced and first particle binding onto G-wires investigated.

Acknowledgements

We thank J. Vesenka, E. Henderson and T. Marsh for introduction into the field and provision of information and unpublished material, S. Diekmann and M. Fändrich for access to the Multimode NanoScope, M. Kittler and A. Csaki for assistance with measurements.
This work was supported by the DFG (FR 1348/3-4) and the Volkswagen Foundation (Priority Area: Physics, Chemistry and Biology with Single Molecules).

REFERENCES

1. Seeman, N. C. Nature 2003,421, 427-31.
2. Yan, H.; Park, S. H.; Finkelstein, G.; Reif, J. H.; and LaBean, T. H. Science 2003,301, 1882-4.
3. Marsh, T. C.; Vesenka, J.; and Henderson, E. Nucleic Acids Research 1995,23, 696-700.
4. Sondermann, A.; Holste, C.; Möller, R.; and Fritzsche, W. (2002) in DNA-Based Molecular Construction (Fritzsche, W., Ed.) pp 103-108, AIP Conference Proceedings 640.
5. Vesenka, J.; Henderson, E.; and Marsh, T. (2002) in DNA-Based Molecular Construction (Fritzsche, W., Ed.) pp 99-109, AIP Conference Proceedings 640.

Hydration Layer Scanning Tunneling Microscopy of "G-wire" DNA

Tamieka Armstrong[a], Jeffrey Root[b], and James Vesenka[a]

[a]*University of New England*
11 Hills Beach Road
Biddeford, ME 04005
Email: jvesenka@une.edu
[b]*California State University Fresno*
2345 E. San Ramon, MS# MH37
Department of Physics,
Fresno, CA 93740-0037

Abstract. Hydration Layer Scanning Tunneling Microscopy (HLSTM) of quadruplex ("G-wire") DNA on mica was carried out under controlled humidity conditions. The G-wires showed remarkable similarity with atomic force microscope images of the same DNA in air, i.e. increased lateral width due to tip broadening but with diameters similar to those measured by x-ray techniques. The G-wire height above the mica substrate and width appeared to decrease slightly with increasing humidity. Though much of the lateral broadening is likely a result of residual buffer salts and the lower resolving ability of HLSTM, the dependence of the DNA height and width on humidity suggests a simple explanation in terms of the hydration layer. An estimate of the increased thickness of the hydration layer of up to 0.6 nm was observed.

INTRODUCTION

G-DNA is a polymorphic family of four-stranded structures containing guanine tetrad motifs (G-quartets) [1,2]. Guanine rich oligonucleotides that are self-complimentary, as found in many telomeric (chromosome ends) repeat sequences, form G-DNA in the presence of monovalent and/or divalent metal cations. The atomic force microscope (AFM) and hydration layer scanning tunneling microscope (HLSTM) are high resolution, near-field, three dimensional imaging devices that were used to explore the structure of linear G-4 polymers at different humidities. These "G-wires" [3] are speculated to form by the self-assembly of the telomeric oligonucleotide sequence d(GGGGTTGGGG) also called d"Tet1.5" monomers (Fig. 1). HLSTM images [5] have suggested that the G-wires are semiconductors [4]. The mechanism of conductivity may be the result of base stacking of G-quartets and caged monovalent cations (Fig. 1). G-wires are well known for their stability when adsorbed to the surface of mica [3,6]. The potential for conduction, stability, uniformity and long lengths make G-wire DNA interesting candidates for molecular wiring [7,8,9]. HLSTM imaging is a sensitive function of both the relative humidity and ionic concentration of the hydration layer (Fig. 2). This study describes a model explaining features of HLSTM images and estimates the thickness of the hydration layer.

CP725, *DNA-Based Molecular Electronics: International Symposium*, edited by W. Fritzsche
© 2004 American Institute of Physics 0-7354-0206-X/04/$22.00

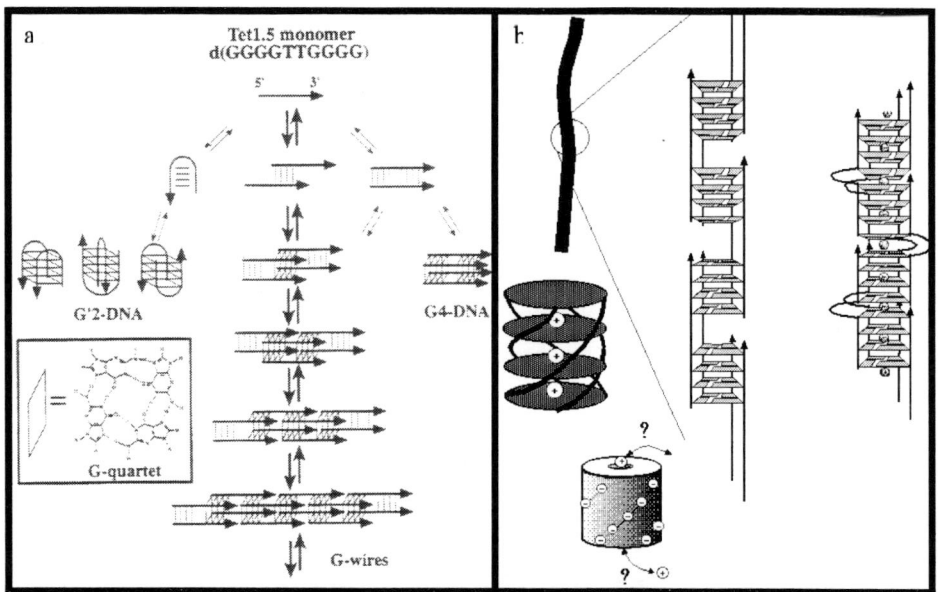

FIGURE 1. The stability of the G-wire DNA is better understood through examination of the hypothesized growth mechanism. The Tet1.5 monomer can form a dimer pair with either a "closed", "looped", or "staggered" conformations as shown in Fig. 1a. In either of the closed or looped conformations no more growth of the G-wires can occur. In the staggered conformation another dimer can attach to the G-wire ladder creating a succession of "sticky ends", enabling multimers to assemble. The process is driven thermodynamically [9]. Monomeric cation species such as potassium or sodium are thought to help stabilize the G-wires down the base-stacked core of the structure as seen in Fig. 1b [3,7]. The thymine groups may act as flexible links that can compress or stretch in solution or after adsorption onto a substrate.

MATERIALS AND METHODS

Quadruplex G-wire DNA was prepared according to the procedure outlined by Marsh et al. [3]. Melting of the $G_4T_2G_4$ monomers (Tet1.5) was maintained with a PCR Thermocylcer (Thermo Hybaid, U.K.), i.e. the growth cocktail and Tet1.5 were is raised to 95°C for ten minutes to promote the melting of fortuitous G-4 structures, i.e. to ensure a monomeric concentration of G-wires. Samples of concentrated G-wires (monomer concentration 1.0 mM) were diluted and incubated for five minutes either directly on freshly cleaved muscovite mica, or on parafilm for ten minutes followed by direct adsorption onto mica at room temperature in a buffer consisting of 10 mM Tris (pH 7.6) and 1 mM $MgCl_2$. The samples were then rinsed with 1 ml de-ionized water to remove excess buffer salts. Freshly prepared G-wire DNA was imaged with a PicoSPM (Molecular Imaging, Tempe AZ) low current STM under controlled humidity with a Digital Instruments (Santa Barbara CA) Nanoscope E controller. G-wires were imaged at RH ≈ 75-85%, bias voltages of -5 to -10 V, and tunneling current of ≈ 1-3.0 picoamperes. A Nanoscope IIIa controller and Multimode AFM operated in Tapping Mode was used to image freshly prepared and dessicated samples.

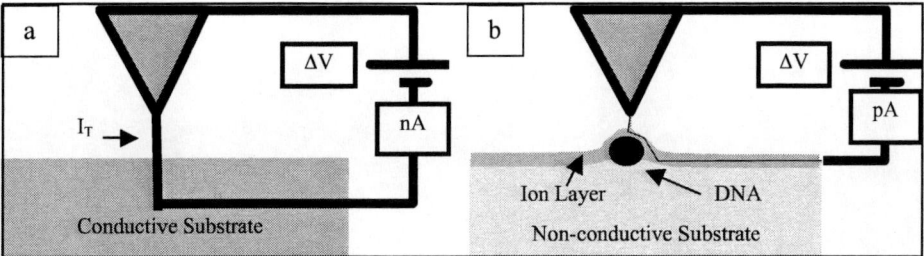

FIGURE 2: In this cross section cartoon of the electron conduction path between tip and sample the traditional STM tunneling current (Fig. 2a) is typically 1000 times greater than the HLSTM current (Fig. 2b). The substantial reduction in HLSTM current is due to the greater resistance of the conduction pathway. The current passes through a thin hydration layer atop the hygroscopic mica surface coated with residual buffer ions in the presence of high humidity. In STM, tunneling currents are typically in the nanoampere range, whereas in HLSTM the tunneling currents are in the picoampere range.

RESULTS AND DISCUSSION

Fig. 3 is an example of a G-wire network that was created by depositing a sample containing a concentrated 24 hour-old G-wire solution onto freshly cleaved mica. The sample is rinsed and immediately imaged in TappingMode™ revealing an oriented network of G-wire strands over the surface of mica in Fig. 3a. The orientation affect is due to G-wire alignment with potassium vacancy sites on the surface of freshly cleaved mica [9]. The density of the G-wire DNA appears to depend upon local variations of the mica surface. For example, imaging a region of the mica surface a millimeter away may provide a distribution of G-wires similar to those seen in Fig. 3c in which the G-wires are clearly separated. In Fig. 3b the same sample from Fig. 3a had been imaged after drying for 24 hours. Note that the G-wire DNA appears much narrower (all three images have the same scale bar) because of the dessication of the hydration layer over the surface. The hydration layer is absolutely essential for imaging of the molecules via HLSTM as seen in Fig. 3c [5].

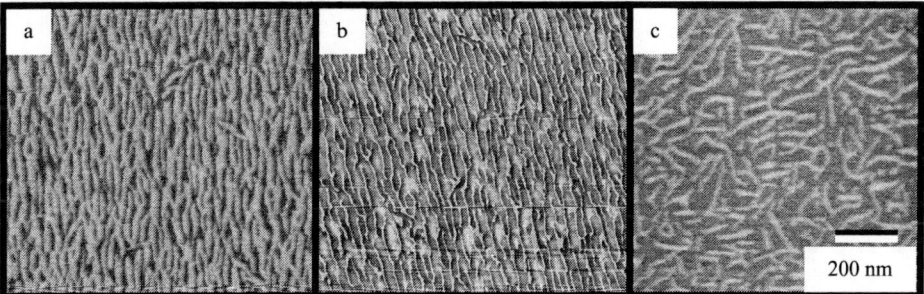

FIGURE 3. G-wires freshly adsorbed onto mica imaged via tapping mode (Fig. 3a) and the same sample imaged by the same tip 24 hours later after drying in an oven at 37°C (Fig. 3b). After drying in an oven the G-wires appear much narrower and the buffer salts appear to distribute themselves in between the DNA strands. It is the freshly made, hydrated form of the G-wires that is essential for HLSTM imaging (Fig. 3c), recorded at 1pA tunneling current, -7V bias and 80% relative humidity in a sealed imaging chamber. Successful HLSTM images of G-wires occur most frequently on freshly prepared samples at high G-wire densities, but NOT G-wire "networks" as seen in Figs. 3a and b. Vertical height range is 10 nm from black to white.

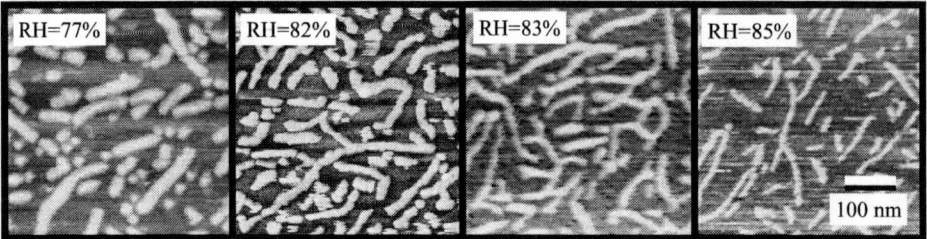

FIGURE 4. Freshly made G-wires imaged by HLSTM at 3pA tunneling current, -7V bias and four different relative humidities in a sealed chamber with different Pt-Ir tips. N.B. the apparent decrease in width as the humidity increases. Vertical height range is 5 nm from black to white.

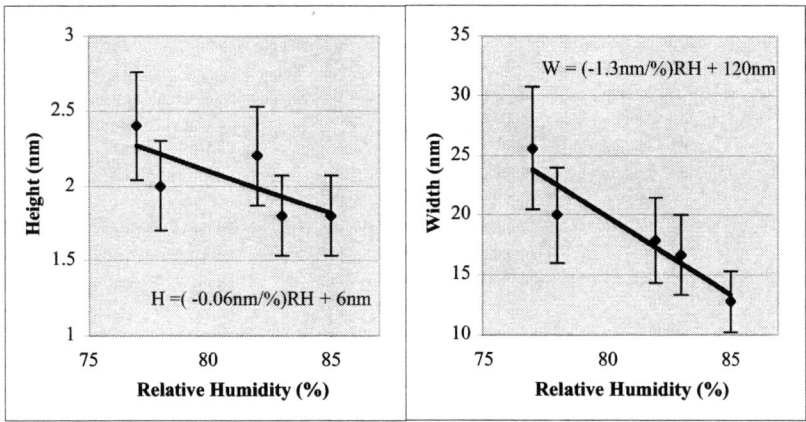

FIGURE 5. Height and width information plotted as a function of relative humidity. The trend lines decrease in both with increasing humidity, but the exact relationship with relative humidity is not possible to establish because of the large error bars and limited range of the humidity measurements. Unlike the decrease in width the slight reduction in height, about 0.6nm, as the humidity increases, is within the measurement error and cannot be established as being significant.

Fig. 4 is a panel representing the humidity dependence of G-wire contrast. Most notable is the apparent reduction in width as the humidity of the imaging chamber was increased, even over a very small range. Fig. 5 reflects the quantitative measurements of the images in Fig. 4, clearly indicating decreasing trend lines. The error bars are so large in the average width measurements, and the relative humidity range so small, it is not possible to glean the exact type of relationship between the two, though it appears the reduction in width is significant. Height information from the same samples indicate a decrease of 0.6 nm as the humidity increased, but the measurement error bars are so large that we are unable to determine if this decrease is significant. In our hands the relative humidity range at which images can be intermittently collected is between 65% and 90%, and only stably between 75% and 85%. Crashing tips is a common casualty as the humidity changes, and all these measurements involved different tips. Consequently the trend lines could represent fortuitous tip broadening as HLSTM *can* involve a slightly different contrast mechanism compared to regular

STM, i.e. through ballistic tunneling [5]. In ballistic tunneling many apical atoms contribute to the tunneling current, reducing the resolution of the image.

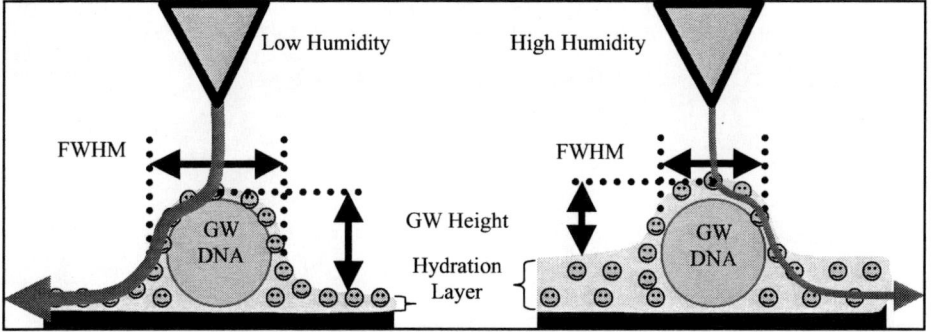

FIGURE 6: The above cross section cartoon speculates about the possible contrast mechanism observed in Fig. 4. The full width half maximum (FWHM) of the G-wire DNA imaged at low humidity will be greater when the hydration layer is thinner than at higher humidity. Also the height of DNA may appear diminished, assuming it is firmly anchored to the substrate, by the rising hydration layer.

The cartoon in Fig. 6 speculates about a possible contrast mechanism that would explain the data in Fig. 5, namely a reduction in height and width of the G-wire DNA as the humidity in the chamber rises. With increased humidity comes greater adsorption of water into the hygroscopic hydration layer on the surface of the sample. If the DNA is firmly anchored onto the surface of the mica, the rising hydration layer would slowly submerge the DNA. The decrease in the height of the DNA is thus reflective of the increase in the thickness of the hydration layer, about 0.6 nm over the range indicated. This thickness is substantially larger than thicknesses measured for water on mica by scanning polarization microscopy [10]. However those results are for pure water without any ions in solution.

ACKNOWLEDGMENTS

Undergraduate research assistant participating in this project: David Bagg, Nicholas Demers, Kristin Eccleston, Matthew Fletcher, Mellissa Holden, Peter Hulsey, Marci Luhrs, and Bethany Rioux. We acknowledge the financial support from the Research Corporation, University of New England, and the National Science Foundation MRI grant DMR-0116398.

REFERENCES

1. Williamson, J.R., Raghuraman, M.K., and Cech, T.R., *Cell*. **59**:871-880 (1989).
2. Williamson, J.R., *Proc. Natl. Acad. Sci. USA*. **90**:3124-3124 (1993).
3. Marsh, T.C., Vesenka, J., and Henderson, E., *Nucleic Acids Res*. **23**:696-700 (1995).
4. Muir, T., Morales, E., Root, J., Kumar, I., Garcia, B., Vellandi, C., Marsh, T., Henderson, E., and Vesenka, J., *J. Vac. Sci. Technol. A*. **16**, 1172-1177 (1998).
5. M. Heim, R. Eschrich, A. Hillebrand, H.F. Knapp, & R. Guckenberger, *J. Vac. Sci. Technol. B* **14**, 1498 (1996).
6. Gottarelli, G., Spada, G.P., and Garbesi, A., in Comprehensive Supramolecular Chemistry, edited by Atwood, J.L., Davies, J.E.D., MacNicol, D.D., and Vögtle, F., Pergamon, New York, 1996, Vol. 9.

7. Rinaldi, R., Branca, E., Cingolani, R., Masiero, S., Spada, G.P., and Gottarelli, G., *App. Phys. Letters.* **78**, 3541-3543, (2001.)
8. Calzolari, A., Di Felice, R., Malinari, E., and Garbesi, A., *Appl. Phys. Letters* **80**, 3331-3334, (2002).
9. Vesenka, J., Marsh, T., and Henderson, E., DNA-Based Molecular Construction, Intern. Workshop on "DNA-based molecular construction", Jena, Germany 2002, Editor: W. Fritzsche, AIP Conference Proceedings 640, New York, 2002, pp. 109-122.
10. Hu, J., Xiao, X.-D., Ogletree, D.F., and Salmeron, M., *Surf. Sci.* **355**, 255 (1996).

DIELECTROPHORETIC MANIPULATION

Stretching DNA
as a template for molecular construction

Masao Washizu[1], Yuji Kimura[1], Takuya Kobayashi[1], Osamu Kurosawa[1,2]
Sayoko Matsumoto[3] and Takayoshi Mamine[3]

[1]Department of Mechanical Engineering, The University of Tokyo
7-3-1 Hongo, Bunkyo-ku, Tokyo, 113-8656, Japan, washizu@washizu.t.u-tokyo.ac.jp
[2]Advance Co.   [3]Sony Corporation

Abstract. The high-specificity self-assembling nature of DNA makes the molecule a candidate for the template for the construction of molecular devices. In order to construct a functional device, the template must be positioned onto a predetermined site on a substrate to allow external connections, and the components must be properly aligned onto the template. A key factor is the high yield of binding, especially when the device consists of many components. Such high yield requires that the bases of the template DNA be exposed so that its counterpart can interact freely, and the template be stretched to avoid folding or coiling that hampers the interaction. We have developed electrokinetic DNA manipulation method, by which a double-stranded DNA is stretched and immobilized bridging over an electrode pair, with the molecular ends anchored while the middle part is left free to interact with foreign molecules. However, double-stranded DNA has a closed structure and the bases inside are not easily accessible. To have a template DNA with accessible bases, two methods are developed in this paper; one being the direct stretching of single-stranded DNA, and the other being the use of a recombination protein to make the bases of double-stranded DNA accessible. The former has an advantage in its simplicity, and the latter in its mechanical stability. We expect that these stretch-and-positioned DNA with accessible base-pairs will lead to the high-yield molecular construction.

INTRODUCTION

The high-specificity self-assembling nature of DNA, together with its properties such as high chemical stability or ease in synthesis, makes the molecule a candidate for the template for the molecular construction [1]. One may conceive an idea of, for example, aligning a molecular device and conducting elements labeled with oligonucleotides, onto a DNA template, as depicted in fig.1. Such alignment scheme should have 1-base (=0.34 nm) resolution. To bring such an idea into reality, there are several technical challenges.

The first is how to place the template molecule on a predetermined location on a substrate (which is between two electrodes in the case of fig.1). The immobilized template should not be entangled but to be maintained straight, to avoid short-circuits, to prevent back-coiling that hampers molecular interaction. The straight conformation should also be preferable in the R&D phase to facilitate evaluation that the elements are in their correct positions.

CP725, *DNA-Based Molecular Electronics: International Symposium*, edited by W. Fritzsche
© 2004 American Institute of Physics 0-7354-0206-X/04/$22.00

The second, which may be more important, is the yield of binding. It's a matter of simple mathematics that, if the probability of successful binding of a component is λ, then the probability for a system consisting of N components will be $\lambda_N = \lambda^N$. Even for the case of $\lambda = 90\%$ (which may be unrealistically high compared with what we observe in bio-chemical experiments), $\lambda_{10} = 0.3$ for a system consisting of 10 components, and for 100 component system, λ_{100} will be as low as 0.00002.

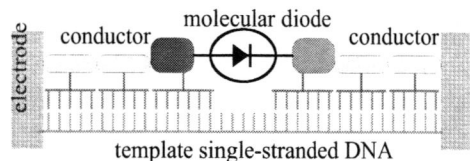

FIGURE 1. An example of DNA based molecular construction.

From this standpoint, such a method as seen in FISH (Fluorescence In Situ Hybridization), i.e. partially denaturing double-stranded DNA and expect the oligonucleotide probe to bind to the opened DNA, will not be suitable, because it has to rely on the chance that the target position is open, and even if the probe once binds, it has to compete with the back-pairing of the original strand.

The requirement for the template therefore will be
  a) Bases must be exposed to allow free interaction with foreign molecules (i.e. elements labeled with oligonucleotides).
  b) The template must be held straight to avoid back-folding.
For this purpose, we have developed two methods for the stretching and immobilization of the template DNA with the bases accessible, which are
  1) direct stretch-and-positioning of single-stranded DNA (hereafter ss-DNA)
  2) stretch-and-positioning of double-stranded DNA (hereafter ds-DNA), in combination with recombination protein which is expected to facilitate sequence-probing and binding to ds-DNA.
The former has the advantage in its simplicity and expected higher yield as the result, and the latter in its mechanical stability.

ELECTROSTATIC STRETCH-AND-POSITIONING OF DS-DNA

We already have a technique for stretch-and-positioning of double-stranded DNA [2-4], whose principle is depicted in fig.2. DNA has the double-stranded helical structure with the diameter of 2 nm and the length of 0.34 nm per base pair (1 μm for 3 kbp). Such a long string-like molecule in water solution takes randomly coiled conformation due to thermal agitation (fig.1 a). When the solution is fed on a pair of electrodes, to which a high frequency (1MHz) voltage up to 1 MV/m (100 V across 100 micron) is applied, DNA polarizes. Due to the interaction between the polarization charge and the external field, electrostatic orientation occurs, and the DNA strand is stretched to a straight shape (fig.1 b). Then dielectrophoresis (DEP) pulls the strand towards the electrode edge where the field is most intense, until one molecular end touches the electrode. The touching end is permanently anchored when electrochemically active metal such as aluminum is used as the electrode material. The mechanism of DNA anchoring onto the electrode edge is possibly electrochemically formed covalent bond between DNA end and aluminum electrode.

The stretching itself is instantaneous, and the whole process completes within a few seconds. We named the process "electrostatic stretch-and-positioning of DNA". The method enables the stretching of individual DNA strands simultaneously, and aligning them with one end in line. The density obtainable with this method is 6 – 7 DNA strands per 1 μm of the electrode contour.

A few advantages for DNA manipulation result from the fact that DEP and electrostatic orientation are effective in a.c. field as well as in d.c. field. By using high frequency ac field, electrolysis at the electrode/water interface can be avoided, so that electrodes can be embedded in water phase. The electrode gap can be made arbitrarily small, so that the high intensity field required for the manipulation of the molecule can be obtained with moderate power supply voltage. By confining the high-field region to a small volume, having large surface-to-volume ratio, heat can efficiently be removed, so a high-intensity field can be created without excessive temperature rise.

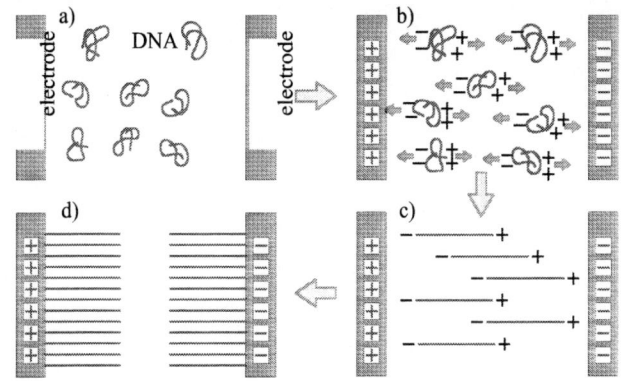

FIGURE 2. Electrostatic stretch-and-positioning of double-stranded DNA

It should be noted that, when electrostatically stretching DNA, there is a practical limit in the medium conductivity. This is partly because to maintain high polarizability of the molecule relative to that of the medium (which is the condition for effective DEP and orientation), partly because to keep temperature rise low.

When a buffer solution of, say >1mM is required, for instance in the case where enzymatic reaction is involved, DNA is stretched and positioned first, voltage is turned off, and then the medium is replaced. This necessitates a DNA anchoring method, to maintain stretched conformation even after the removal of the field. The permanent anchoring onto aluminum electrode edge can conveniently be used for this purpose.

Fig.3 depicts the electrode system to obtain stretched DNA in field-free condition [4], which we named DNA high-wire system. It consists of a pair of energizing electrodes on a glass substrate, and a few thin strips of aluminum having no electrical connection, which we call floating-potential electrodes (FPE). The spacing between the floating-potential electrodes are made slightly smaller than the length of DNA to be immobilized, and the glass surface between the electrodes are etched down about 1μm. When the outer electrodes are energized, DNA is stretched and pulled towards the edge of the electrodes. When one molecular end of DNA touches an electrode, it is anchored. At this moment, the other end extends to the edge of adjacent electrode, and is dielectrophoretically pulled-in to be anchored. Because the glass surface is lower, DNA is held free except at both ends, and thus free interaction with foreign molecules is guaranteed. The function of FPE here is to deform electrostatic field to create field maxima for DEP trapping of DNA. These electrodes are better left to the floating potential: when electrodes are connected to power supply with low impedance, charge injection creates jet-like flow at the very vicinity of the electrode edge, which hampers the approaching of DNA ends.

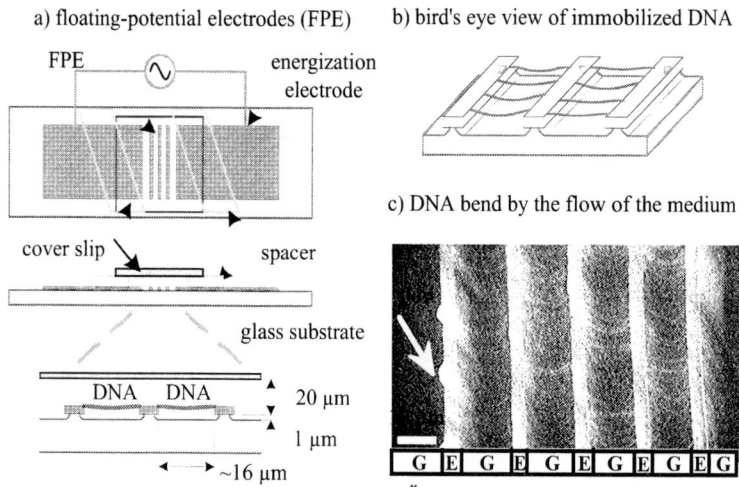

FIGURE 3. DNA high-wire system

ELECTROSTATIC STRETCH-AND-POSITIONING OF SS-DNA

The mechanism of DNA polarization is the counter-ion polarization [5]. DNA has negatively charged phosphate backbones, which attracts positive ions in solution to form counter-ion cloud. When the field is applied, the counter-ion moves along the strand, and excess positive ion is accumulated at the downstream side of the field, while negative backbone charge is exposed at the upstream side, hence the polarization occurs.

ss-DNA has negatively charged backbone, half the density of ds-DNA, so it may as well be stretched in electrostatic field. However, stretching ss-DNA was found to be not as easy as that of ds-DNA. ds-DNA is rather stiff, due to the paired structure. On the other hand, in ss-DNA, bases can rotate around the backbone, and may form internal pairs or folding. It is very flexible molecule. In addition, ss-DNA is mechanically weak, even a single defect in the backbone results in disintegration.

Another technical problem is that there is no good fluorescence tag for staining ss-DNA, in contrast to ds-DNA to which bright intercalating dyes are available.

So, we started with a development of fluorescence-tagged ss-DNA as shown in fig.4. The principle is to use PCR to synthesize ss-DNA (fig.4 a), during which process fluorescence-labeled monomer (fluorescein d-UTP) is incorporated (fig.4 b). By denaturing thus synthesized ds-DNA, we can obtain fluorescence-labeled ss-DNA, however, it is the mixture of the two strands. Even if the stretching in electrostatic field is observed with the sample, we may be observing annealed ds-DNA which is more easily stretched. Hence, we have to completely remove one of the strands to avoid such back-pairing. To do so, one of the primers is biotinylated as shown in fig.3 a. After PCR, the synthesized ds-DNA is attached to avidin-labeled magnetic micro particles (fig.4 c), denatured in 0.1M NaOH, and the strands in the supernatant are recovered (fig.4 d). The purity of thus obtained ss-DNA is checked by adding S1 nuclease that digests ss-DNA. The maximum length of fluorescence-labeled DNA obtained by this method with a reasonable yield was about 10 kb.

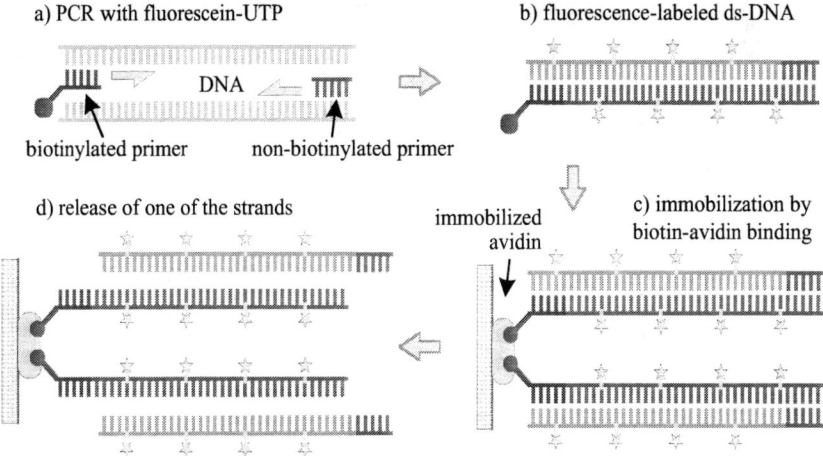

FIGURE 4. Fluorescence labeling of single-stranded DNA

By using the fluorescence-labeled ss-DNA thus obtained, we observed its stretching under electrostatic field. First, comparison of stretched length is made between ds- and ss-DNA for various field intensities, which is plotted in fig.5. This measurement is made for DNA on the floating electrodes to avoid hydrodynamic perturbations.

For the case of ds-DNA (fig.5 a), with the increase in applied field (direction shown by an arrow in the figure), the stretched length increases up to c.a. 3.4 µm, which is in agreement with the length calculated from the structural constant 0.34 nm per 1 base. When the field is decreased, the stretched length shrinks reversibly, following the same curve as when the field is increased.

On the other hand, for the case of ss-DNA (fig.5 b), it shows a clear hysterisis; it requires higher field for stretching, but requires less field for maintaining the stretched shape. This is attributable to the interaction inside the strand, partial base-pairing or hydrophobic nature, that requires more energy to separate. The maximum stretched length of ss-DNA is observed to be only 2/3 of that of ds-DNA. If ss-DNA is stretched until its backbone is straight, it should have the length twice that of ds-DNA. Why the stretched length of ss-DNA is even shorter than that of ds-DNA is still an open question. It may be that the stretching force is still insufficient to break remaining internal base-pairs, or may be due to base stacking.

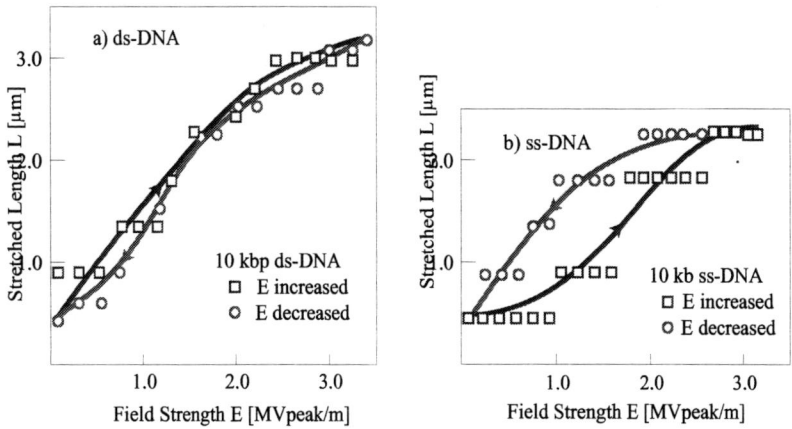

FIGURE 5. Electrostatic field strength v.s. stretched length for ds- and ss-DNA

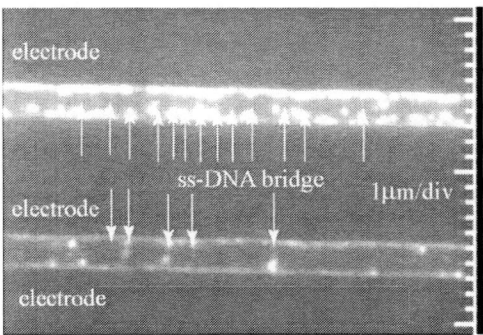

FIGURE 6. Stretched single-stranded DNA anchored at both ends onto the micro electrode

Fig.6 shows a photo of ss-DNA successfully immobilized bridging over floating electrodes (the bright spots on the electrode edge, in particular at upper electrode gap, are aggregates of ss-DNA which is attracted there by dielectrophoresis). In actual experiment, bending of the immobilized ss-DNA due to flow is observed, proving the double-ended anchoring is achieved with the middle part freely suspended. We are now observing the binding of oligonucleotide probes to the stretch-and-positioned ss-DNA.

BASE-PAIR FORMATION WITH THE AID OF
RECOMBINATION PROTEIN

As a method to create the template structure for molecular constructions, we have shown the stretch-and-positioning of ss-DNA. However, a problem with ss-DNA is that it is mechanically weak. Even single breakage on its backbone should result in disintegration. If bases of double-stranded DNA can be exposed and made accessible, we will have more stable system. We are investigating the use of recombination protein for this purpose.

Recombination protein, such as RecA or Rad51, takes part in homologous recombination. Fig.7 schematically depicts the function of RecA. A diploid has two sets of almost identical gene. When one of the genes is damaged, repair process is initiated (fig.7 a). The strands near the damaged point are digested by an exonuclease for a certain length, and single-strands are exposed (fig.7 b). RecA binds there (fig.7 c), and goes into the other set of intact DNA, to look for the homologous sequence (fig.7 d). When the sequence is found, DNA is reconnected, and repaired (fig.7 e). This means that ss-DNA/RecA complex has the function of sequence-probing of, and binding to, already existing base pairs of ds-DNA. It is equivalent to having a base-exposed ds-DNA, to which foreign ss-DNA can bind. The above picture is for RecA, which has an affinity to ss-DNA. On the other hand, DmcI, another recombination protein, is known to have larger affinity to ds-DNA rather than ss-DNA. Although the detail is still to be clarified, it suggests the possibility that it has a function of actually exposing the bases of ds-DNA.

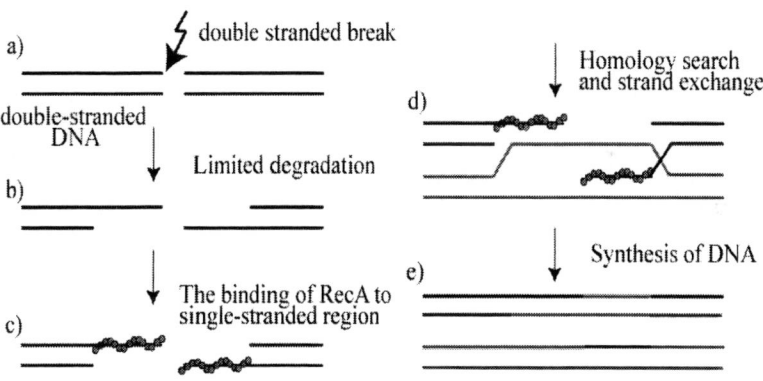

FIGURE 7. Base probing during homologous recombination

From this point of view, we are making single-molecule observation of the interaction between ss-DNA/RecA complex and ds-DNA, using DNA high-wire system of fig.3. In order to make a dynamic observation, the complex must be fluorescence-tagged. This is done by using partially single-stranded DNA. 23kb ds-DNA is digested by Exonuclease III from ends to make single-stranded part of c.a. 600 bases. Then it is cut at 4.6 kb from one end by a restriction enzyme Nru I, which leaves a blunt end. Thus, c.a.18kb double-stranded DNA having 600 base single-strand extension at one end is obtained (we hereafter call it probe DNA). When intercalator dye is added, it binds to the relatively long double-stranded part to make the probe DNA bright enough to be observed in single-molecule level under a fluorescence microscope, while the single-stranded part remains unaffected.

The fluorescence-labeled probe DNA is mixed with RecA, and fed by a gentle flow to the DNA high-wire system where ds-DNA (λ-DNA, 48kb = 16μm in length) is suspended. Fig.8 is a photograph of the ss-DNA/RecA complex mounting on a ds-DNA. It should be noted that, because the substrate between the electrodes are etched down (fig.3), a complex on the substrate surface will be out of the focal plane, i.e. the visible white dot in the photo must be on the ds-DNA bridge. In the experiment of fig.8, sliding diffusional motion of the complex is observed, back and forth along the ds-DNA for a distance as long as 5 μm. But such long-distance movement is relatively rare. In most case, when the complex carried by the intentional slow flow of the medium hits ds-DNA bridge, it binds near the contact point.

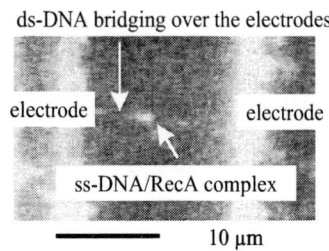

ds-DNA bridging over the electrodes

electrode   electrode

ss-DNA/RecA complex

————— 10 μm

FIGURE 8. Sliding of single-stranded DNA/RecA complex along a double-stranded DNA

In fig.9 is shown the observed final binding position of the complex as a function of normalized position. In this experiment, the sequence homologous to that of the probe DNA is almost at the center. It is seen in the figure that there is a peak near the homologous sequence, and dips on both sides. Also seen is the decay at both extremes of the ds-DNA bridge.

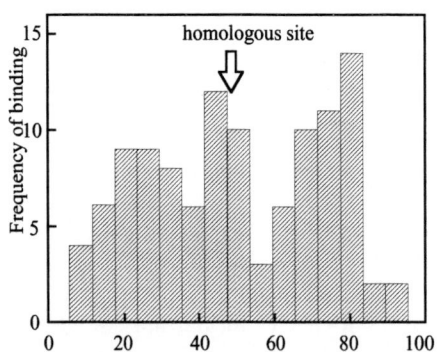

Normalized position of the DNA probe along the target DNA [%]

FIGURE 9. Histogram of binding position of ss-DNA/RecA complex on ds-DNA high-wire

This is still preliminary data and subject to further investigations, but our tentative interpretation is as follows. As has been proved for DNA-interacting proteins, such as RNA polymerase [6] or restriction enzyme [7], sequence-probing function is associated with a sliding diffusional motion along DNA. By first weakly and non-specifically binding to anywhere on the DNA strand and sliding along it, a protein can reach to its target sequence far easily compared with just relying on random 3-dimensional diffusion. And when the sequence is found during the sliding, the specific and stronger binding occurs.

From this point of view, if the non-specific binding is stronger than as it should be, the probing molecule will not be able to diffuse for an adequate distance to reach its destination, i.e. the target sequence. This seems to be what we observe in fig.9. If the DNA probe, carried by flow or diffusion, happens to hit near the destination, it can diffuse and finally reach there. This is the reason why fig.9 has a small peak at the target sequence and small dips around it, where the width of the dip is the measure of possible diffusion length. On the other hand, if the probe happen to hit a distant place, it will never reach to its destination, and stops anywhere with equal probability. The probes hitting near the extremes of DNA bridge can reach to the molecular end, and may be released or adsorbed on the metallic electrodes, thus fig.9 has lower frequency at the template ends.

This hypothesis suggests that there is optimal non-specific interaction strength. The DNA probe without RecA does not interact with the DNA bridge because of electrostatic repulsion of phosphate backbones, so the cause of non-specific binding must be due to the charge on RecA. The DNA probe in the experiment might have been too heavily loaded with RecA. We are now trying to locate the optimal condition for non-specific binding.

CONCLUSIONS

For the practical use of DNA as the template for molecular construction, high yield of base-pair formation must be guaranteed. To achieve this goal, we identified two factors, a) to have a template DNA strands with the bases exposed to allow free interaction with foreign DNA bases, and b) to have a stretched template to avoid folding that makes bases less accessible. Based on the consideration, two methods has been developed to obtain stretch-and-positioned DNA, which are

1) single-stranded DNA high-wire system based on the electrostatic stretch-and-positioning. A hysterisis in the stretching field strength v.s. stretched length is observed, and the maximum stretched length of ss-DNA obtained is about 2/3 of that of ds-DNA.

2) the use of recombination protein RecA to make the bases of ds-DNA accessible to foreign oligonucleotide. It is found that controlling the strength of non-specific binding, which determines the distance of sliding diffusional motion, is a crucial factor.

The former has an advantage in its simplicity, and the latter in the mechanical stability. These methods now allow us to quantitatively evaluate the binding yield, and to find out and improve the conditions for the molecular construction.

ACKNOWLEDGMENTS

The author would like to thank Dr. Hiroyuki Kabata of Kyoto Univ., Dr. Takashi Kinebuchi of RIKEN, and Prof. Hitoshi Kurumizaka of Waseda University for collaborations and discussions. This work is in part supported by BRAIN (Seiken-Kiko) Research and Development Program for New Bio-industry Initiatives, the Ministry of Education (Kakenhi A14205037), Advance Co. and Sony Corporation.

REFERENCES

1. Fritzsche, W. ed.: "DNA-Based Molecular Construction", American Institute of Physics Conference Proceedings 640 (2003)
2. Washizu, M. and Kurosawa, O: "Electrostatic Manipulation of DNA in Microfabricated Structures", IEEE Trans. IA, Vol.26, No.6, p.1165-1172 (1990)
3. Washizu, M., Kurosawa, O, Arai, I., Suzuki, S. and Shimamoto, N.: "Applications of Electrostatic Stretch-and-positioning of DNA", IEEE Transaction IA. Vol.31, No.3, p.447-456 (1995)
4. Yamamoto, T., Kurosawa, O., Kabata, H., Shimamoto, N. and Washizu, M.: "Molecular surgery of DNA based on electrostatic micromanipulation", IEEE Transaction IA, Vol.36, No.4, p.1010-1017 (2000)
5. Suzuki, S., Yamanashi, T, Tazawa, S., Kurosawa, O and Washizu, M: "Quantitative analysis on electrostatic orientation of DNA in stationary AC electric field using fluorescence anisotropy", IEEE Transaction IA, Vol.34, No.1, p.75-83 (1998)
6. Kabata, H., Kurosawa, O., Arai, I., Washizu, M., Margarson, S.A., Glass, R.E. and Shimamoto, N.: "Visualization of single molecules of RNA polymerase sliding along DNA", Science, Vol.262, p.1561-1563 (1993)
7. Kabata, H., Okada, W. and Washizu, M.: "Single-Molecule Dynamics of the Eco RI Enzyme using Stretched DNA: Its Application to In Situ Sliding Assay and Optical DNA Mapping", Jpn. J. Appl. Phys. Vol.39, p.7164-7171 (2000)

Monitoring Dielectrophoretic Collection of DNA by Impedance Measurement

R. Hölzel and F. F. Bier

Fraunhofer Institute for Biomedical Engineering, Molecular Bioanalytics & Bioelectronics, Arthur-Scheunert-Allee 114-116, Bergholz-Rehbrücke, 14558 Nuthetal, Germany

Abstract. Double stranded M13 phagemid DNA has been locally concentrated between interdigitated electrodes by dielectrophoresis. RF electric fields at 0.1 MHz and 1 MHz have been applied with field strengths exceeding 1 MV/m. Impedance changes were monitored by analysis of the driving signals or with an additional sensing signal applied at 1 kHz. DNA collection was found to be reflected clearly by changes in the capacitive part of the setup's impedance. This presents a new method for a label free, purely electronic detection of macromolecules.

INTRODUCTION

For the production of electronic circuits on a molecular level a well controlled spatial manipulation of the building blocks is necessary. Positioning can be accomplished on a real single molecule basis, e.g. by atomic force microscopy (AFM). This approach is only feasible in the laboratory, but not for a future series or mass production, although there are examples of parallelization of AFM by the use of several cantilevers simultaneously [1, 2]. Alignment of potential electronic building blocks like DNA and nanotubes can be achieved by molecular combing applying a hydrodynamic fluid flow [3, 4] and by dielectrophoresis (DEP) using alternating current (AC) electric fields [5-9]. Whilst molecular combing is achieved relatively easily over tens of micrometres distances, it is difficult to build structures that are more complicated than parallel stripes and grids [3]. On the other hand positioning and alignment of DNA by AC electric fields can be controlled rather easily within dimensions starting from tens of nanometres. Complex patterns can be formed depending on the shape, number and excitation scheme of the electrode arrangement. While there is a lot of both experimental and theoretical work on the AC electrokinetic properties of biological cells [10-15], there is no systematic study of AC electrokinetic effects on molecules. Still, only a fundamental understanding of the mechanisms underlying molecular DEP will allow a full exploitation in nanotechnology. Such studies are hindered by the experimental problem of quantifying the dielectrophoretic response of molecules. Usually molecules are observed by fluorescence microscopy. For this purpose they have to be labeled by a fluorescent marker. Depending mainly on the detector's properties the temporal resolution is limited and quantitative results suffer from bleaching of the fluorophores. Therefore a label free and possibly direct electronic detection scheme for macromolecules would be quite advantageous. Here we describe a system which measures the local concentration of DNA by electrical impedance changes.

CP725, *DNA-Based Molecular Electronics: International Symposium*, edited by W. Fritzsche
© 2004 American Institute of Physics 0-7354-0206-X/04/$22.00

COMBINATION OF DIELECTROPHORESIS AND IMPEDANCE MEASUREMENTS

Material and Methods

Dielectrophoresis and other AC electrokinetic effects are widely used for the characterization and separation of biological cells [16-18]. Usually the objects' kinetic response to the field action is determined by microscopical observation. An alternative. approach has been introduced by two groups which quantify the cell concentration in the close vicinity of the DEP electrodes by impedance measurements [19, 20]. Here, the particles that have been concentrated preferentially at the electrode edges contribute to the overall impedance of the setup to a larger degree than those dispersed in the bulk solution (Fig. 1). By this the electrical impedance is a measure of the particle concentration close to the electrodes and hence a measure of the dielectrophoretic response.

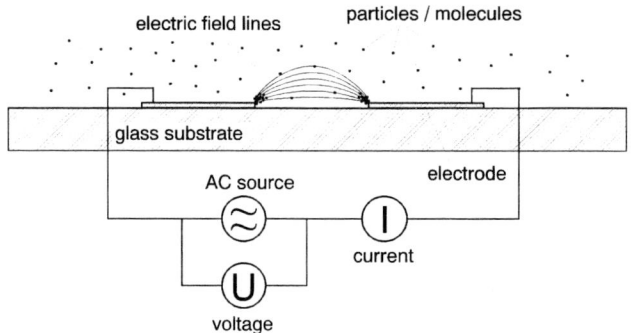

FIGURE 1. Combination of dielectrophoresis and impedance measurement. Due to the inhomogeneity of the electric field particles and macromolecules experience a force towards the electrode edges. There they lead to a change in the field induced current which can be measured.

Determination of the impedance can be performed with the help of an oscilloscope comparing phase and amplitude of the AC voltage and current, respectively [20]. In this straight forward approach the same field is used for both DEP collection and impedance monitoring.

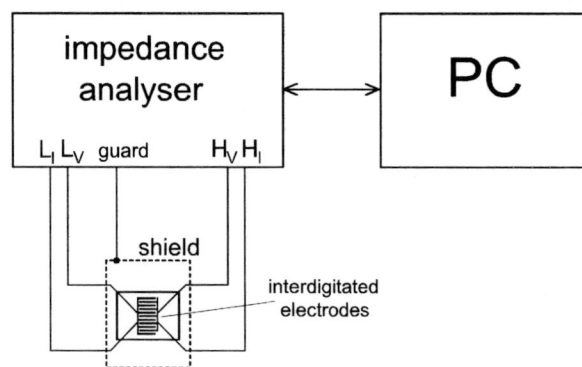

FIGURE 2. Experimental setup. The impedance analyser serves both as DEP field source and as measurement device. A personal computer controls the analyser and records the data.

In order to change the dielectrophoretic and the detection field independently Allsopp et al. [21] used two signal sources. For maximal sensitivity at minimal electrode polarisation they chose 800 Hz for measuring the impedance. DEP excitation frequency was varied from 1 kHz to 50 MHz. Both signals were coupled by a transformer. Phase and amplitude at 800 Hz were determined through a lock-in amplifier. Concentration changes were found to be equally reflected by both phase and amplitude.

In this study an impedance meter (Hioki 3532) has been used for delivering the driving voltage for dielectrophoresis as well as for determining impedance changes (Fig. 2). The instrument automatically measures amplitude and phase of voltage and current, respectively. From these it derives circuit parameters like impedance, phase, capacitance or conductance. Interfacing to a personal computer allows automatic control of the device's settings and registration of measured data.

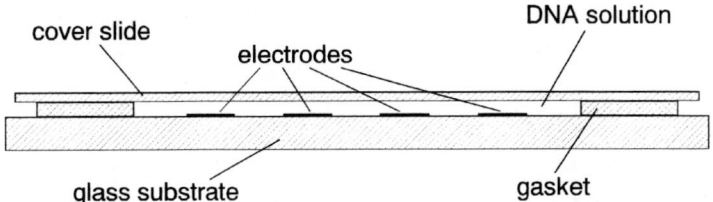

FIGURE 3. Measuring chamber. A few µl DNA solution are placed onto the electrodes and are secured against evaporation by ring of adhesive tape and a cover slide.

Interdigitated electrodes were produced from gold on a glass substrate by standard photolithography. Each of the comblike electrodes consists of 50 parallel lines, each 2 µm wide and 100 µm long, with a gap width of 2 µm. The surface is protected by a 100 nm thick layer of SiO_2/Si_3N_4. The chip is placed on a copper cladded epoxy glass circuit board. The bonding pads are connected to the board's circuit paths by thin wires and electrically conductive glue. The fluid to be placed onto the electrodes is confined by double sided adhesive tape and a cover slide allowing microscopical observation (Fig. 3).

In most of the studies dealing with the concentration of DNA by AC electrokinetic effects frequencies around 1 MHz have been applied at field strengths around 1 MV/m [5, 22-25]. Therefore, similar values were chosen as a starting point in this work, only at somewhat stronger fields (1.0 MHz, 5 V_{RMS}). Double stranded DNA from a linearized M13 phagemid at a concentration of 20 nM was used. M13 comprises about 7250 base pairs corresponding to a length, if stretched, of 2.5 µm, which can bridge the electrode gap.

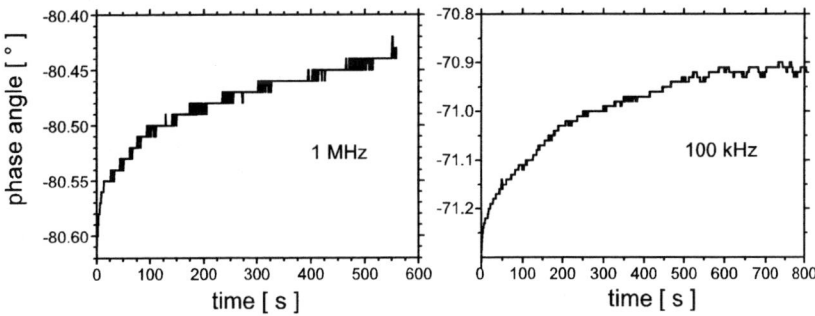

FIGURE 4. Time course of the phase shift between voltage and current of the dielectrophoresis signal at 1 MHz and 100 kHz, respectively.

Experimental results

A small exponential increase in phase angle was found (Fig. 4. a) with a time constant of 190 s. In a similar study using fluorescence microscopy [26] a time constant of 24 min was found at the same frequency, but at lower field strength (2 V over a 1.7 μm gap). Assuming a square dependence of DEP on field strength, as it is well proven and understood for particles bigger than some μm [11, 13, 27], yields a time constant of 240 s, which agrees quite well with this work. However, such a square dependence still remains to be proven for the dielectrophoresis of molecules. The value of $|Z|$ was found to decrease linearly by 0.6% within 10 min. This presumably is a consequence of the very small chamber volume (about 1 μl) leading to some evaporation and hence an increase in ion concentration. Applying 100 kHz led to a similar behaviour (Fig. 4 b) resulting in a slightly higher time constant of 230 s.

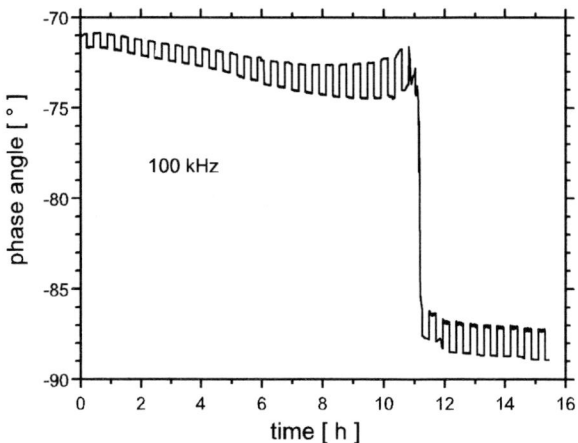

FIGURE 5. Time course of the phase angle during dielectrophoretic collection at 100 kHz. Field amplitude was alternating between 5 V and 0.1 V every 14 minutes.

This experiment has been carried out for nearly 16 hours with the signal voltage being alternated between 5 V and 0.1 V (Fig. 5). The steplike phase change with lower phase angles at the lower excitation amplitude is believed not to be a result of a true concentration change, since this should exhibit an exponential dependence. This effect might be caused by electrode polarization, however, it remains to be investigated further. The exponential increase during the first 5 V application exactly continues after the break, when applying only 0.1 V.

This means that the dielectrophoretically attracted DNA adheres to the SiO_2/Si_3N_4 surface, which had been cleaned before with 3M NaOH. The long term drift of the phase can be explained by evaporation and temperature changes. The large drop to almost 90° phase angle after 11 hours is a consequence of the fluid drop withdrawing from the electrodes. 90° would represent a purely capacitive behaviour with no resistive part, i.e. air being above the electrodes.

In a further experiment the time scale of amplitude change was decreased to 37 s (Fig. 6). Again the nearly constant phase angle at 5 V and at 0.1 V, respectively, is obvious. This time an "overshoot" at each beginning of the 5 V period is discernible. Averaging of these 23 periods and fitting to an exponential function yields a time constant of this overshoot's decay of 1.7 s. In this experiment the electrode surface had been cleaned only by rinsing with ethanol and distilled water. Thus this overshoot might indicate a partial release of some regions of the DNA strand in contrast to a complete adhesion to an NaOH treated surface.

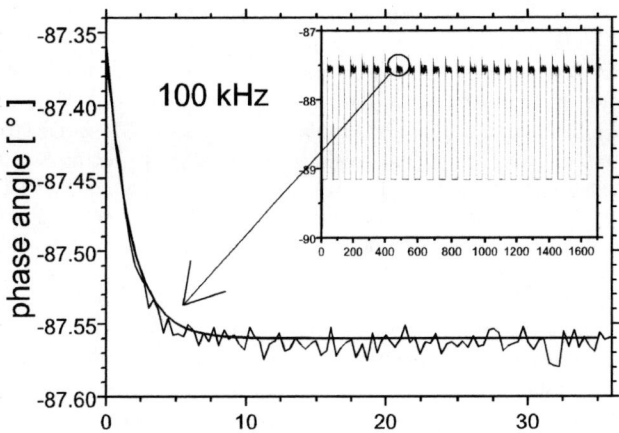

FIGURE 6. Time course of the phase angle during dielectrophoretic collection at 100 kHz. Field amplitude was alternating between 5 V and 0.1 V every 37 s (see insert). 23 consecutive "overshoots have been averaged. Data have been fitted with a first order exponential decay.

From the observation that under the present circumstances DNA mostly adheres to the surface after being attracted it follows that short interruptions in DEP field application do not change the overall properties of the system. This can be exploited to increase sensitivity by using different frequencies for DEP and for detection as it has been done by Allsopp and coworkers [21]. For this purpose these authors used two signal sources, a lock-in amplifier and a coupling circuit.

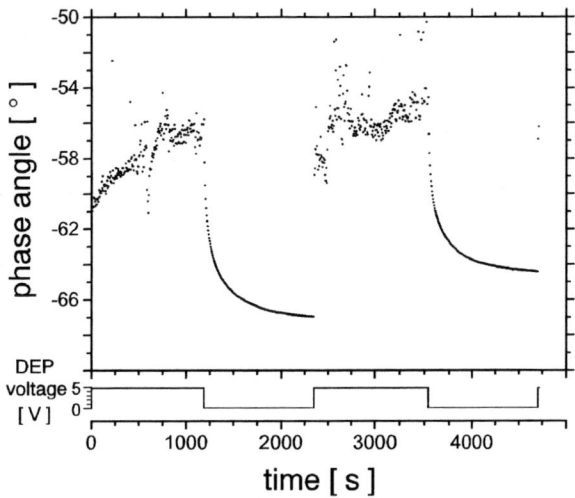

FIGURE 7. Time course of the sensing field's phase angle. Monitoring signal was at 1 kHz and 0.1 V. Dielectrophoresis voltage (lower trace) changed between 0.1 V and 4.9 V every 20 minutes.

A similar result can be achieved using only the impedance analyser. This is done by interrupting dielectrophoresis by short periods of a sensing signal differing in voltage and amplitude. For this purpose DEP was carried out at 100 kHz with the amplitude switching from 4.9 V to 0.1 V every 20 minutes. After every 4 seconds the signal was changed to 1 kHz at 0.1 V for about a second. According to [21] the use of 1 kHz leads to a relatively high sensitivity in low conductive media at acceptable electrode polarization. The result (Fig. 7) shows an exponential phase angle change of the sensing signal which is more than one order of magnitude larger than the change of the actual DEP signal. At the same time the phase angle has become much more "noisy". This probably reflects the DNA's DEP behaviour being observable now in more detail. Yet, this might also be caused by artifacts introduced by the measuring principle itself. It has been found that the field response of DNA is an order of magnitude stronger in the audio frequency range [7, 25, 28] than around 1 MHz. Thus the sensing field might introduce significant dielectrophoretic or even electrophoretic effects despite amounting to only 2% of the DEP field's strength. It is known that high frequency fields also stretch DNA molecules in parallel to the field lines [6, 9, 29], whilst low frequencies just lead to an attraction towards the electrode edges. Thus, alternating between high and low frequencies may repeatedly reorganize the DNA strands' alignment and by this lead to stronger impedance fluctuations. The increased time constant of about 500 s as compared to 230 s from Fig. 4 b is partly a consequence of the DEP field being activated for only about 80% of the time. The phase angle's decay time of 220 s during field off times is surprisingly long. Again this might be a consequence of the sensing field still being present. Interestingly the signal is much less "noisy" during field off periods which also points to an interference of low frequency and high frequency effects.

CONCLUSION

It has been shown for the first time that the AC electrokinetic response of macromolecules can be determined by purely electronic means. This is even possible with a rather simple setup using only an impedance meter as the electronic part. As compared to fluorescence microscopy there are no interfering effects like photobleaching. Therefore a systematic, quantitative investigation of molecular dielectrophoresis will be possible much easier than with conventional, fluorescence based methods. Work is in progress aimed at an increased sensitivity allowing the analyte's concentration to be reduced, which might finally lead to an electronic device for the trapping and sensing of a single molecule.

ACKNOWLEDGEMENT

We are indebted to X. Marschan for preparing the DNA. Funding of this work by the European Union within the project Nanocell is gratefully acknowledged.

REFERENCES

1. P. Vettiger, G. Cross, M. Despont, U. Drechsler, U. Durig, B. Gotsmann, W. Haberle, M. A. Lantz, H. E. Rothuizen, R. Stutz, and G. K. Binnig, *IEEE Trans. Nanotechnol.* **1**, 39-55 (2002).
2. S. Kramer, R. R. Fuierer, and C. B. Gorman, *Chem. Rev.* **103**, 4367-4418 (2003).
3. Z. Deng and C. Mao, *Nano Lett.* **3**, 1545-1548 (2003).
4. A. Bensimon, A. Simon, A. Chiffaudel, V. Croquette, F. Heslot, and D. Bensimon, *Science* **265**, 2096-2098 (1994).
5. M. Washizu and O. Kurosawa, *IEEE Trans. Ind. Appl.* **26**, 1165-1172 (1990).
6. H. Kabata, O. Kurosawa, I. Arai, M. Washizu, S. A. Margarson, R. E. Glass, and N. Shimamoto, *Science* **262**, 1561-1563 (1993).
7. C. L. Asbury and G. van den Engh, *Biophys. J.* **74**, 1024-1030 (1998).
8. R. Krupke, F. Hennrich, H. B. Weber, M. M. Kappes, and H. von Löhneysen, *Nano Lett.* **3**, 1019-1023 (2003).
9. R. Hölzel and F. F. Bier, *IEE Proc.-Nanobiotechnol.* **150**, 47-53 (2003).
10. G. Fuhr, U. Zimmermann, and S. G. Shirley, " Cell Motion in Time-varying Fields: Principles and Potential" in *Electromanipulation of Cells*, edited by U. Zimmermann and G. A. Neil, CRC Press, Boca Raton, 1996, pp. 259-328.
11. X.-B. Wang, Y. Huang, R. Hölzel, J. P. H. Burt, and R. Pethig, *J. Phys. D: Appl. Phys.* **26**, 312-322 (1993).
12. X.-B. Wang, Y. Huang, F. F. Becker, and P. R. C. Gascoyne, *J. Phys. D: Appl. Phys.* **27**, 1571-1574 (1994).

13. J. Gimsa and D. Wachner, *Biophys. J.* **75**, 1107-1116 (1998).
14. R. Hölzel, *Biochim. Biophys. Acta* **1450**, 53-60 (1999).
15. R. Pethig, V. Bressler, C. Carswell-Crumpton, Y. Chen, L. Foster-Haje, M. E. Garcia-Ojeda, R. S. Lee, G. M. Lock, M. S. Talary, and K. M. Tate, *Electrophoresis* **23**, 2057-2063 (2002).
16. F. F. Becker, X. B. Wang, Y. Huang, R. Pethig, J. Vykoukal, and P. R. C. Gascoyne, *Proc. Natl. Acad. Sci. USA* **92**, 860-864 (1995).
17. P. R. C. Gascoyne and J. Vykoukal, *Electrophoresis.* **23**, 1973-1983 (2002).
18. G. Fuhr and C. Reichle, *Trends Anal. Chem.* **19**, 402-409 (2000).
19. K. R. Milner, A. P. Brown, D. W. E. Allsopp, and W. B. Betts, *Electron. Lett.* **34**, 66-67 (1998).
20. J. Suehiro, R. Yatsunami, R. Hamada, and M. Hara, *J. Phys. D: Appl. Phys.* **32**, 2814-2820 (1999).
21. D. W. E. Allsopp, K. R. Milner, A. P. Brown, and W. B. Betts, *J. Phys. D: Appl. Phys.* **32**, 1066-1074 (1999).
22. F. Dewarrat, M. Calame, and C. Schönenberger, *Single Mol.* **3**, 189-193 (2002).
23. V. Namasivayam, R. G. Larson, D. T. Burke, and M. A. Burns, *Anal. Chem.* **74**, 3378-3385 (2002).
24. R. Hölzel, N. Gajovic-Eichelmann, and F. F. Bier, *Biosens. Bioelectron.* **18**, 555-564 (2003).
25. R. Hölzel and F. F. Bier, *IEE Proc.-Nanobiotechnol.* **150**, 47-53 (2003).
26. R. Hölzel, *J. Electrostat.* **56**, 435-447 (2002).
27. H. A. Pohl, *Dielectrophoresis*, Cambridge University Press, Cambridge, 1978.
28. C.-F. Chou, J. O. Tegenfeldt, O. Bakajin, S. S. Chan, E. C. Cox, N. Darnton, T. Duke, and R. H. Austin, *Biophys. J.* **83**, 2170-2179 (2002).
29. W. A. Germishuizen, C. Wälti, R. Wirtz, M. B. Johnston, M. Pepper, A. G. Davies, and A. P. J. Middelberg, *Nanotechnol.* **14**, 896-902 (2003).

Manipulation of metal nanoparticles in micrometer electrode gaps by dielectrophoresis

Robert Kretschmer, Wolfgang Fritzsche

Institute for Physical High Technology, POB 100239, 07702 Jena, Germany

Abstract. The integration of molecular structures into microscopic electrode arrays can be achieved by dielectrophoresis of gold nanoparticles in electrode gaps. Using microelectrodes realized by photolithography, we demonstrate here the generation of pearl chain arrangements of nanoparticles in structures accessible for standard technologies. In order to preserve the individual particle structures in the final nanowire arrangement, various strategies were employed. This method for defined positioning of nanoparticle chains offers the potential to wire DNA or DNA superstructures after conjugation to the particles and dielectrophoresis. It allows a parallel processing of molecular structures and their integration into microsystem technology.

INTRODUCTION

Based on their unique electronic and optical properties, metal nanoparticles emerged as an interesting part of today's molecular nanotechnology. Due to a highly established bioconjugation methodology in microscopy, their integration with the molecular framework of molecular nanotechnology is straightforward. Moreover, they exhibit the potential to introduce novel physical properties into molecular constructs, thereby enhancing the potential of this field. The problem is the defined integration into e.g. electronic setups, such as a two-terminal contact structure. The use of electrostatic effects could overcome the manipulation problem of nanoparticles. Using miniaturized electrodes, particles can be arranged in the gap using the electric field. The literature presents two approaches that are differentiated by the applied electrode gap and the preservation of the original particle integrity. The first approach uses gaps in the lower nanometer range, usually fabricated by e-beam techniques. These gaps allow the electrostatic trapping of individual particles with an extremely precise localization [1,2]. However, the fabrication of the gaps is no standard technique, and the characterization difficult. On the other side, electrode gaps accessible by parallel techniques such as photolithography or even precision mechanics with gap sizes in the micro to millimeter range have been reported. Applying an AC field results in the growth of nanoparticle chains bridging the gaps due to dielectrophoresis [3]. In this case, the structure in the gap is usually less defined regarding the width (more than one particle) and the integrity of individual particles (often melted into a rather homogeneous mass) [4,5]. This paper addresses a combination of the advantages of both approaches: The highly defined single file of particles from the small gaps, and the ease of fabrication from the larger gaps. This goal is realized by the optimization of the parameters for dielectrophoresis in micro gaps.

CP725, *DNA-Based Molecular Electronics: International Symposium*, edited by W. Fritzsche
© 2004 American Institute of Physics 0-7354-0206-X/04/$22.00

MATERIALS & METHODS

Standard photolithographic techniques were applied to fabricate the microelectrodes of 100 nm gold with a titanium adhesion layer on oxidized silicon substrates. Droplets of 5 µl of gold nanoparticles (5-30 nm diameter, British Biocell International, Cardiff, UK) were applied on the electrode gap. A frequency generator Voltcraft FG-506 (Conrad Electronics, Hirschau, Germany) applied an ac-potential of 3-8 V and 0.02-1.00 MHz through wire probes. The shortcut nanowire current was limited by a series resistor of 0.5 to 10.5 kOhm.

The ultramicroscopic characterization was conducted using a scanning electron microscope JSM 6700 F (Jeol, Zaventem, Belgium).

RESULTS & DISCUSSIONS

The aim of this work is to study whether metal nanoparticles can be used as molecular handles to access nanoscale constructs such as DNA conjugates etc. Therefore, nanoparticles should mediate between the macroscopic world of measurement technique (via microscopic planar electrode array) and the molecular constructs. This vision could be realized with methods of construct conjugation to nanoparticles prior to nanoparticle alignment into single-file chains. Because the conjugation of (bio) molecules with metal nanoparticles has a long and successful tradition, we focus here on the second part. As discussed in the introductory section, the positioning of nanoparticles in electrode gaps by electrical fields is a known technique, but the published procedures could not combine ease of use (photolithographic electrodes) with preservation of the nanoparticle integrity (usually a melting is observed for microelectrodes).

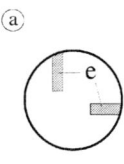 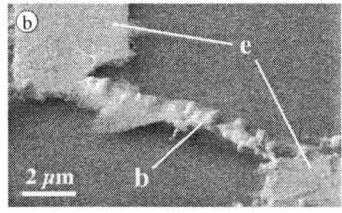

FIGURE 1: Dielectrophoresis of metal nanoparticles in a microelectrode gap. a) Scheme of electrode geometry with a gap width of several micrometer. b) SEM of such an electrode arrangement after dielectrophoresis of 30 nm nanoparticles. A bridge (*b*) is visible connecting both electrodes (*e*). The globular substructure of the bridge is too large for nanoparticles and probably reflecting melted aggregates of material.

In order to adapt the principle of dielectrophoresis onto our system, preliminary experiments were carried out using electrodes with a gap width of several micrometer (cf. Scheme in Fig. 1a). Typically it took more than 5 min application of 16 V before a sudden drop in voltage indicated bridge formation. An ultramicroscopic characterization revealed that the two electrodes (top left and lower right in Fig. 1b) were connected by a bridge (center). Although some globular structures are visible, there is no clear substructure of the bridge pointing to the used 30 nm nanoparticles. We assume that after shortcut the current heated the particle bridge and induced a melting. This hypothesis would explain the larger globular structures as generated by the nanoparticles and/or the electrode material. Moreover, the upper electrode is apparently damaged by this procedure, another indication for the occurring damaging forces.

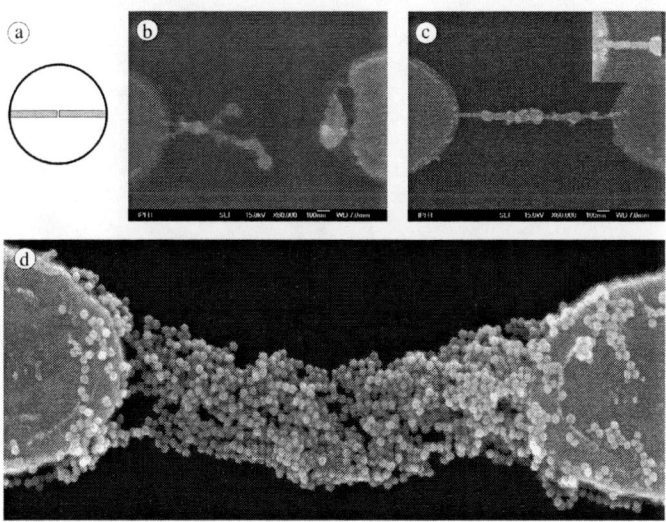

FIGURE 2: Dielectrophoresis of nanoparticles in a 1 μm electrode gap. a) Scheme of electrode geometry. b) Experiment with 5 nm nanoparticles. c) Single-chain bridge yielded using 30 nm particles. A zoom of the left region of the bridge (inset) shows that the nanoparticles are apparently partially melted together. d) Wider bridge of 30 nm particles, still preserved in their original shape.

In order to limit these damages, a decrease in voltage would be an option. The experiments showed that the height of the voltage determined the time required for bridge generation. Higher voltages resulted in decreased time needed for bridge formation. A decrease in voltage leads thereby to extended growth times, and even to no growth at all. This effect can be counteracted by smaller gaps. So we changed to a new electrode design with a gap size of only 1 μm (scheme in Fig. 2a).

We studied the effect of particle size, using diameters of 5-30 nm. The largest particles yielded the best and most reproducible results. With smaller particles (such as 5 nm), it was more difficult (and often impossible) to achieve gap closure (cf. Fig. 2b). Using 30 nm particles, bridges with widths ranging from one particle diameter (Fig. 2c) to more than ten particle diameters (Fig. 2d) were observed. A closer inspection of the narrow sections of the bridges revealed that in the case of a single connection (Fig. 2c) the particles are at least partially melted (cf. zoom), but in cases with several connections (Fig. 2d) the individual particles in these regions are still recognizable.

Although we used now an electronic setup to turn off the current immediately after closing the gap, the current still seems to have an effect on the particle chains. In the case of just one connection, this chain acts like a fuse and melts at least partially. In wider bridges, several chains contribute to the conduction, and every single chain experiences less current and thereby less damage.

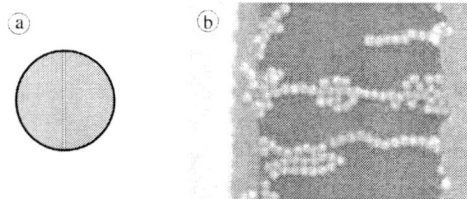

FIGURE 3: Experiments with very wide electrodes (resulting in an extended gap). a) Scheme of the used geometry. b) SEM of a part of the gap after the dielectrophoresis shows the chain-like arrangement of the particles and the high degree of preservation.

We applied another electrode geometry, using very wide electrodes that should induce the generation of multiple chains (Fig. 3a) in order to provide enough cross-sectional area to preserve the particles. The result of this experiment is shown in Fig. 3b: Several individual bridges could be observed in the electrode gap, and all particles are nicely preserved.

Conclusions

These results present a proof of concept for the realization of well-preserved nanoparticle chains in micrometer electrode gaps. Further steps will include the inclusion of (bio) molecules, probably as self-assembly monolayer, in order to characterize e.g. the electrical properties of such layers.

Acknowledgements

We thank F. Jahn for SEM imaging, M. Urban and V. Baier for helpful discussions.

REFERENCES

1. Bezryadin, A.; Dekker, C.; and Schmid, G. Applied Physics Letters 1997,71, 1273-1275.
2. Amlani, I.; Rawlett, A. M.; Nagahara, L. A.; and Tsui, R. K. Applied Physics Letters 2002,80, 2761-2763.
3. Pohl, H. A. (1978) Dielectrophoresis, Cambridge University Press, Cambridge.
4. Khondaker, S. I.; and Yao, Z. Applied Physics Letters 2002,81, 4613-4615.
5. Hermanson, K. D.; Lumsdon, S. O.; Williams, J. P.; Kaler, E. W.; and Velev, O. D. Science 2001,294, 1082-6.

APPENDICES

PROGRAM (1)

Thursday, May 13

20:00 Optical Museum   (Downtown, 10 min from Hotel IBIS)   Guided Tour & Come-Together Bufet

Friday, May 14

08:15 Pick-up from Hotel IBIS (please meet at front desk)

08:30 Registration

09:00 Opening (IPHT at the Campus Beutenberg)

DNA Nanotechnology

09:10   N. Seeman (New York)
Structural DNA-Nanotechnology
09:35   I. Willner (Jerusalem)
*Metallic and Conductive-Polymer Nanowires on
DNA Templates for Electronic Applications*
10:00   A. Woolley (B.-Young Univ.)
*DNA-templated fabrication of carbon nanotube
and metal nanowires*
10:25   A. Filoramo (Gif-sur-Yvette)
*Non-covalent binding of DNA to carbon nanotubes
controlled by biological recognition complexes*

10:50   ---- coffee break / posters ----

DNA Manipulation

11:20   W. Fritzsche (Jena)
*A parallel approach to DNA nanotechnology: Step-
by-step self assembly of metal nanostructures*
11:45   M. Washizu (Tokyo)
Stretching DNA as a template for molecular construction
12:10   M. Heller (San Diego)
*Electric Field Based Process for Producing Linear
Electronic/Photonic Transfer Nanostructures*
12:35   A. Estabrook (Santa Barbara)
*Enabling methods for nucleic acid and
nanoparticle-based molecular electronics*

13:00   ---- lunch break / posters ----

DNA Conductivity : Mechanism

13: 40   R. Di Felice (Modena)
*Theoretical investigation of metal-containing
DNA-based helices*
14:05   F. Gervasio (Lugano)
Charge transfer and oxidative damage in DNA fibers
14:30   B. Giese (Basel)
Chemistry At A Distance: Electron Transport Through DNA

PROGRAM (2)

Friday, May 14 (cont.)

14:55   A. Rakitin (Toronto)
*DNA Electrical Properties: Natural Variety
and Experimental Limits*

15:20   ---- coffee break / posters ----

DNA Conductivity: Experiments

15:45   A. Kasumov (RIKEN)
Conductivity and Induced Superconductivity in DNA
16:10   C. Nogues (Weizmann Inst.)
*Electrical properties of DNA characterized by
conducting-atomic force microscopy*
16:35   D. Porath (Jerusalem)
*SPM and Charge Transport Measurements of
Various DNA and DNA-based Molecules*
17:00   J. Gomez (Madrid)
*AFM based techniques to measure conductivity
in DNA based Nanowires*

17:30   ---- Excursion ----

19:00   ---- Thuringian Bratwurst Dinner ----

Saturday, May 15

DNA Construction I

08:30   H. Yan (Duke)
New Structures for DNA-Based Nanofabrication
08:55   M. Helfrich (Penn State)
*DNA nanoparticles: Towards Deterministic
Bottom-up Assembly of Metal Nanoparticles*
09:20   K. Mir (Oxford)
*Spatially addressable self-assembly, combing and
nanoparticle binding of single DNA polymers*
09:45   T. Kawai (Osaka)
DNA based electric and magnetic devices

10:10   ---- coffee break / posters ----

DNA Construction II / Technical Developments

10:30   J. Vesenka (Univ. New England)
Auto-orientation of "G-wire" DNA
10:55   K. Williams (Delft)
*Assembly of Nanotube-based Electronic Devices
by Biomolecular Recognition*
11:20   J. Schlütter (LOT Darmstadt)
*Dip Pen Nanolithography: A new tool
for the fabrication of nanostructures*

11:45   ---- end of the scientific program ----
12:00   ---- lunch ----

PROGRAM (3)

Poster Presentations

Monitoring dielectrophoretic collection of DNA by imped. spectroscopy
F. Bier (Potsdam)

Oligofunctional DNA-Au Nanoparticles for the Assembly of Particle Layers
B. Ceyhan (Dortmund)

A toolbox for molecular construction based on DNA,
metal nanoparticles and microstructured surfaces
A. Csaki (Jena)

Formation and electrical properties of DNA-immobilized
gold nanoparticle monolayers
E. Hesse (Aachen)

Coupling G-wires to metal nanoparticles
C. Holste (Jena)

High Resolution DNA Imaging by Atomic Force Microscopy
D. Klinov (Moscow)

Pearl chain formation of nanoparticles in microelectrode gaps
by dielectrophoresis
R. Kretschmer (Jena)

Adsorption of DNA molecule and DNA Patterning on Si substrate
S. Tanaka (Osaka)

Generation of gold nanoparticles in micro reactors
J. Wagner (Ilmenau)

SYMPOSIUM NOTES

by James Vesenka

Ned Seeman (New York University) – Structural DNA-Nanotechnology
e-mail: ned.seeman@nyu.edu
DNA is the genetic material of all living organisms (Picture of Eve). DNA basics: problem of DNA is that it is helical. Easy to design immobile branched junctions, to minimize sequence symmetry, thus not strictly linear. Sticky end cohesion: smaller affinity, hydrogen bonding, ligation. Sticky ends are structural well defined. Central concept of structural nanotechnology: combine branched with sticky ends to make nanotechnology. High resolution/structural DNA used as bricks and mortar. Low resolution/compositional DNA used as mortar only (includes nanoparticles or long range ordering.) What is the intellectual goal of structural DNA nanotechnology? Controlling the structure of matter in 3D. Objectives: Architectural control and scaffolding for nanomechanicsl devices. A new suggestion for producing macromolecular crystals: DNA box-like species as a host for guests. A method for organizing nanoelectronic components. Append DNA with macromolecules. Polyhedra catenanes: Cube, truncated octahedron. Construction of crystalline arrays is required for lattice design components, predictable interactions, and structural integrity. Marshmallow impaled with uncooked rotini. DX Isomers, double crossover DNA molecules. DPE, DAE, DPON DPOW, DAE+J. 2D DX Arrays, schematic of lattice containing 1DX tile and 1 DX+J Tile, matrix every 32 nanometers, seen in AFM. DX Triangle Arrays, simple bulged 3-arm junction triangle (1996), doubled the thickness got a pseudo-hexagonal array, nice honeycombed structures, great AFM images. Progress toward three-Dimensional arrays: A 3D trigonal DX Lattice. X-ray arrays supported the triangle structure, but only 10 A data. Made big crystals. Organizing other Species with DNA. Prototyping the control of molecular topology. Nucleic acid knot (NA) with pendent groups, pendent groups linked, polymer released from nucleic acid. Individual components, nylon coming off the DNA. Mass spec indicates success. Organizing 1.4-nm gold nanoparticles: imaged by STEM with. Organizing 10-nm gold nanoparticles. DNA parallelogram with gold particles at each corner. DNA nanomechanical devices. B->Z DNA transition. Sequence dependent device, machine cycle of the PX-JX$_2$. System to test the device checked with AFM. Can see parallel and zigzag. DNA walking biped, have the flexible DNA walk over rigid DNA structure. Possible crosslinking products. Denaturing gel showing crosslinking results for each step. Can see the strands change as the DNA walks. Summary of results: What's next? Listed scientific challenges for this field in 2002 during last conference. Quite a few of the challenges have been successful. Organization challenges for this field: Promote the study of the control of organization of matter, to unite component disciplines of the field, increase prominence. Founding

international society for nanoscale science computation and engineering, June 11, 2004, Milano Italy.

Itamar Willner (Hebrew University, Jerusalem) Metallic and Conductive-Polymer Nanowires on DNA Templates for Electronic Applications:
e-mail: willnea@vms.huji.ac.il

DNA as a template for the construction of Au-nanowires: DNA functionalized semiconductor nanoparticles, polyaniline-DNA conductive wires, DNAzmes as active components for the generation of polymer and metallic wires. Actin as a template for the generation of wires. Why DNA? E.g. synthetic construction, availability of enzymes for scission/elongation, and polymerization. dsDNA provides sequence-specific intercalators or the association of cations into the phosphate units. Sequence addressable binding of proteins to NA provides means for patterning. dsDNA, intercalate gold nanoparticles, and image via light activation. Hard to get uniform coverage of the DNA. The idea is to use the particles to make wire. Use of cancer cells to identify telomeric DNA. Use the telomeres as a template for wires. Telomerase to grow dNTPs/dUTP-NH$_2$, intercalate gold particles and can secure μm long nanowires. Or, just telomere to makes ssDNA, intercalate through complimentary double strand assembly with nanoparticle, also provides nice wires. Three-arm branched junction with amine branched DNA, delivers nice "Y" shaped gold intercalated wires. We can also combine three DNA arms at the center. Metallization is relatively easy. Metal semiconductor hybrid devices: attached cadmium sulfide nanoparticles with complementary strands linked to CdS. Evaluated using absorption spectroscopy. Photocurrent increases with number of layers, but DNA has no effect on conductivity. Must decorate DNA with positively charged particles for charge transport. Photocurrent jumps with ruthenium oxide. How do we make contacts? Semiconductor nanoparticle, cadmium sulfide, attached dye labeled dUTP+dNTPs, to examine fluorescence energy transfer, find it improves transfer. Semiconductor with two gold contacts on either end: polyanaline-DNA conductive wire. ds-β DNA, attached copolymer of analine such that FAD attached to DNA backbone. Glucose oxidase was attached to the wire to make electron transfer from DNA to gold wire. G-quadruplex with amine intercalated can result in chemiluminescence repeats on the surface. Amplified the system to get gold attached to G-wire templates. Can amplify detection of DNA. Actin as a template for the generation of wires: motor protein, coupled to myosin, and ATP, myosin starts to move. Took actin monomers to polymerize to f-actin, modify with gold NP, decorated filament, depolymerize, repolymerize and enlarge with gold, 2-μm long, 80-nm high nanowires. Attached unlabeled filaments and got a "bar of gold" with unlabeled f-actin at edges. Can go from non-patterned actin to decorated to make monomer gold contacts. Tested gold decorated f-actin for electronic conduction and measured a highly conductive nanowire. In principle a motor with gold to get a nano transporter, to glass slide with myosin, followed spectroscopically, add ATP, see four pictures at 5 second intervals. Can calculate motility rate, 254nm/s. Have motor that moves on a surface.

Adam Woolley (Brigham Young University) DNA-templated fabrication of carbon nanotube and metal nanowires:

e-mail: atw@byu.edu

Background: A tale of two Y's. In 1910 students tried to construct a 200m tall school logo on the adjacent hill. The logistics were very challenging as the aspect ratio had to be compensated for line of sight. In 2004 we are trying to construct a DNA nano-Y, 20nm across: DNA templated nanocircuits. DNA as a nanofabrication template, 2nm, linear, aligned, specificity (base-pairing) conductivity? Background: negative DNA charge electrostatically attracts cations: reduce to make silver metal nanostructures, 100nm diameter wires. Objectives: probe the early stages of metal deposition, explore nanowire synthesis with metals and carbon nanotubes, overcome specificity and template cleavage problems. Explore more complex DNA-metal assemblies. Macroscale process, nanoscale alignment, uses molecular combing. Clean off silicon surfaces, modify with poly-lysine (1ppm), DNA droplet translation, image with AFM, example was circular single stranded DNA. DNA templated synthesis of Cu nanowires. Treat DNA with Cu^{2+}, interacts electrostatically, add ascorbic acid, reduced to Cu metal. Cu metal on DNA is nucleation site for more deposition. Height increases to 20-Å, but only 50% coated with Cu. Nonspecific nanoparticle background. Metallized DNA surfaces: reducing nonspecific background, followed by reduction, causes the background. Try to block nonspecific adsorption sites, with high surface affinity cation. Align DNA, treat with cesium, then add Ag, Cs reduces background of Ag nanowires. Examine backscattered electrons in SEM. Generalized with dsDNA, metallized without K^+, Cu nanorods made using K^+ treatment. Cu nanowire fabrication causes DNA fragmentation. Aqueous Cu^{2+} makes hydroxyl radical which cleaves DNA. Got an aprotic solvent to reduce OH^-, less cleavage. To increase Cu, Ag density, use a low dielectric solvent. Dimethyl sulfoxide makes greater length, surface masking, length 7-μm long, 7-nm tall. Localization: DNA to surface, then CNT to DNA. DNA is hydrophilic. Use 1 PMA (pyrenemethylamine) aggregate on DNA to get specific localizations of SWNT, 60% were aligned on DNA, covering only 1% of surface. Only 5% of total DNA length is decorated with SWNT. Evaluation of nanowires: Are these DNA templated nanostructures really nanowires? Useful to probe same structure at various stages of synthesis, need to measure electrical characteristic, grow nanowires across electrode arrays, must be planar. Have aligned DNA across electrodes, done metallization, preliminary conductivity measurements, resistance is ~ Gβ . Looking to e-beam lithography to go to smaller spaces. Fabrication of more complex architectures: Can we make more complex DNA nanostructure designs? AFM: native DNA, TEM-Ag metallized DNA, optimistic about moving onto more complicated devices.

Arianne Filoramo (Gif-sur-Yvette CEA Caclay, ENS Paris and Motorola Labs) – Non-covalent binding of DNA to carbon nanotubes controlled by biological recognition complexes:

e-mail: filoramo@drecam.saclay.cea.fr

Introduction: Integrated circuit scaling trends and roadmaps, Moore's Law and Density of silicon devices. Motivation: problem to scale down to sub 10-nm is costly. Seek alternatives to traditional electronics. 1959 Feynam: "There is plenty of room at the bottom", 2004 "which one is technology is most interesting." Alternative candidates to CMOS, issues for sub 10-nm scaling: sensitive to variation of their physical dimension, power/energy dissipation, speed, cost. What's needed to go beyond? Bottom-up construction using self-assembly, multi-disciplinary expertise, new system architecture. Molecular and nanoscale electronics: Using DNA strands as a scaffold and self-recognition, decide to work with carbon nanotubes (CNT). CNT have exceptional properties, well characterized mechanical, capillary, and electrical properties. Different types of devices make, junctions, gates, all fabricated by random depositions. Have to first find with AFM, how to make compatible with large-scale fabrications? Selective placement via surface preparation and manipulation. Selective attachment, lithography, resist protection/clean, stick patch formation with gaseous APTS deposition, lift-off resists, exposed to CNT solution, yield of deposition of 100% on APTS, 100 nm wide patterns. Next step is to connect CNT to allow fabrication of integrated devices. Put gates on nanotube after deposition. Still requires lithography. Can use DNA to contract. DNA-SWNT: covalent or non-covalent chemistry. Perfect nanotube is not possible to perform via covalent chemistry: COOH/COO⁻ defects are needed. Strong acid treatments are required to obtain this chemistry. Sonicate for several hours, possibly with ill effects such as lowering transport performances. E.g. Raman spectroscopy, SP_2 to SP_3 transitions appear to have been effected. Nanotubes are very sensitive to their environment. Examples of chemically modified characteristics: Transport and optical absorption can completely change for better or worse (mostly worse). Functionalization of CNT by DNA strands e.g. through streptavidin (SA) and biotin complex. SA has hydrophobic interactions with CNT, AFM shows decoration with SA bumps. Then Biotin-DNA will attach to tubes, attached to APTS treated mica. Unfortunately the DNA can attach anywhere. 10kB DNA was used in the AFM example shown here. When DNA is not biotinilated DNA does not bind. Even with strong mechanical action, like washing, anchorage persists. Depositions followed by combing of DNA on surface gives alignment. Double biotin DNA can bind to same carbon nanotube. Studied DNA conductivity and in our experiments it does not appear to be conductive. So we chose DNA metallization. Similar results to Mertig et al., unfortunately nanotube metallizes as well. RecA polymerization, challenge is to go beyond standard lithography using the DNA to localize the CNT.

Wolfgang Fritzsche: A parallel approach to DNA nanotechnology: Step-by-step self-assembly of metal nanostructures
e-mail: fritzsche@ipht-jena.de
A requirement for further development in DNA-based nanoelectronics is access via 2 or 4-way device using a parallel approach. Why? Statistics needed in establishment phase, and technology application. Examined metal enhanced nanoparticle DNA. Also looked at long DNA molecules metalized with a nanoparticle as a tunneling barrier in between. Two years ago we used a flow through chamber to orient the DNA. This was complicated because of the functionalization process and to get DNA over the contacts resulting in low contact efficiency. Recently we have used an

improved technique, molecular combing through capillary action into a regular orientation with parallel alignment. The combing is clearly seen with fluorescence imaging and when the DNA is decorated with nanoparticles. Electrodes are positioned under drops of DNA before combing, appeared to get DNA to align with electrodes and span the electrode gaps. Deposition of some crud occurred as well. Orientation seen in fluorescence, AFM, and SEM (when attached with charged nanoparticles.) DNA terminated by nanoparticles is easy, harder to get gold in spaces between. Very low efficiency. So, decorate by non-specific binding through positively-charged gold and enhance with silver. Beads on a string in AFM, and enhanced by silver can be seen by optical techniques. Modification of DNA aligned with electrodes, still not very uniform, looked at I-V curves providing >72 GW resistance.

Masao Washizu: Stretching DNA as a template for molecular construction.
e-mail: washizu@washizu.t.u-tokyo.ac.jp
Template molecular base pairs: a possible application for molecular devices or more complicated structures, but is such a device realistic? Clearly they must provide a straight template to avoid entanglement, facilitate evaluation, avoid back-coiling, and need high yield of binding: E.g. $\beta_{total} = \beta^N$, if N=100, β=99%, then $\beta_{total} = 0.37$. FISH (fluorescence in-situ hybridization), fluorescence dye intercalated in DNA, but is a low yield process. In order to achieve high yield binding, exposes bases (ssDNA) with stretched shape to allow for binding. Electrostatic stretch and positioning of dsDNA. Apply an electric field (1MV/m), DNA moves toward electrode until one end touches and permanently anchors on aluminum electrode. Would like to keep stretched after removing the field. Make electrode gap equal to stretched length of DNA. DNA is immobilized ABOVE the solid surface of the DNA. Images are all fluorescent. Why ssDNA more difficult to stretch? Ds-DNA stiffness due to paired structure and charge backbone, with persistent length (L_p) of 300 bp, ssDNA has hydrophobic folding, internal base pairing and L_p of 10-20 bp, and harder to fluorescence label ssDNA. Stretched length as a function of applied filed, some hysteresis, but about 0.6mm/MV/m for ds DNA, 1mm/MV/m for ssDNA. Successfully stretch ssDNA bridging over the electrode gap, not successful with complimentary stabilization. ssDNA is molecularly weak. Perhaps some protein to keep dsDNA open then open sides can be reacted with. We tried homologous recombination and so we sought complimentary fluorescence labeling. Frequency of binding from homologous site, did not get very specific binding, with slight peak at homologous site but non specifically bound elsewhere. Video shows fluorophore drifting along the DNA. Need to control strength of non-specific binding.

Michael Heller (UC San Diego) – Electric Field Based Process of Producing Linear Electronic/Photonic Transfer Nanostructures.
e-mail: mjheller@ucsd.edu
Silicon microelectronics are already into the 100-nm range. Are there any advantages to smaller structures, will it be non-CMOS? For true micro/nano/molecular devices we need heterogeneous integration process. Lithographic process has been pressed

close to its limits and will provide nanoscale CMOS semiconductors. Truly heterogeneous integration of photonic microelectronics is difficult and expensive, many nanoscale devices exist, but none integrated into higher order 2D/3D structures. Self-assembly is valuable, but how do we integrate these structures. Looking for higher order photonic and electric transfer properties in self-assembled DNA structures. E.g. donor/acceptor for fluorescence energy transfer. Works in first order transfers, but higher order affects does not happen (originally discovered in 1984 while working at Standard Oil of Indiana). Today, pulsing quantum dots for rapid sensitive and selective DNA genotyping and infectious agent detection. Looking at linear DNA quantum dot chains. Hit and miss process, difficult for getting a viable device. Bionanotechnology, magnetic tags, gold particles, clever cantilever, nano-bar codes, but not cost effective. Need precision nanostructures and better nanofabrication processes. Development of DNA chips, cartridge device, research/diagnostic system (Nanogen), use electric field to orient molecules in the development of top-down and bottom-up processes for nanofabrication and heterogeneous integration. E.g. electric field DNA transport and concentration: electric field control of DNA hybridization, i.e. directed self-assembly! Highly parallel electric field process for cell or particle separation. Highly parallel fashion, electric field directed assembly of microspheres and nanospheres, 3D assembly by covering layer after layer. Devices and arrays for electric field directed heterogeneous assembly. Heterogeneous integration processes for micro/nanofabrication – synergy of top-down with bottom-up processes. Electric field directed was proven in the fabrication of precision DNA nanoparticles (US patent # 6652808). Paradigm change – eliminate the $Billion Fabrication". Put a hetero-fabrication unit in your hand, make your own integrated electronic/photonic/MEMS devices. Missing area of bionanotechnology. Nanorotors based on ATP synthase: Mechanism for chemical to mechanical energy conversation. What is the take home message regarding bio-inspired nanotechnology? Concepts of self-assembly are inaccurate and apply only to homogenous and first order processes, high field recognition of molecules.

August Estabrook (UC Santa Barbara) Enabling methods for nucleic acid and nanoparticle-based molecular electronics.

e-mail: aestabrook@chem.ucsb.edu

Collaboration between Riech (biochem.)/Strouse (chem.)/Cleland (physics). The big picture: nanoparticle/DNA biomaterials are ideal building blocks for nano-scale circuit assembly. Our goal is to develop enabling recombinant DNA and nanoparticle synthesis and electrical characterization of elementary architectures. Gold nanoparticles, synthesized in house with weak ligand shells, can be recapped by strong binding ligands and allow chemistries to be modulated. A novel construct this chemistry has enables is the synthesis of water-soluble semi-conducting nanoparticles (ZnS over CdSe). These nanoparticles have band gaps, can be used as gates, are tunable, and have unique electrical properties. Surface ligands can be heterogeneous, modulate solubility, and facilitate coupling. Directed assembly of nanoparticles on DNA scaffolds: thiolated at both 5' ends of PCR DNA. Electrostatics - positively charged gold couples to negatively charged DNA. Amine, incorporation of modified dNTPs. Seeking to demonstrate a high amount of control. Interested in electrostatics so examined DNA on SiO^2, thiolcholine. Increasing density of 13-nm gold. See lots more non-specific annealing, but also packed tighter together. Have done molecular combing with 4-nm charged gold particles. Making wires. Use thiol chemistry to attach DNA scaffold to gold electrodes. Been able to get large bundles of DNA, I-V curves in the 10^{12}-β region. 300-nm gap and found no conductance of naked DNA. Coupled with amines, through the incorporation of modified bases into the PCR products: dATP, dGTP, dCTP (NO dTTP) + dUTP amino allyl, nanoparticle position is now controlled by DNA sequence. Asymmetric PCR makes only one strand at a time, gold is only on one side. Examining DNA connects such as cruciform and zinc finger: highly characterized protein sequence that can bind pairs of DNA. Made the double headed zinc finger, seen with a gel, but not with the AFM. Dissociation of one head messes with the association of the other. Creating large overhangs, used 20+ bp overhangs, 1) RNA primed PCR, 2) abasic primed PCR, 3) re-anneal. In all cases we added thiolated gold.

Rosa Di Felice (INFM National Center for nanostructures and BioSystem at Surfaces, Modena) Theoretical investigation of metal-coating DNA-based helices
e-mail: rosa@unimore.it
Why exploring charge carriers in metal-containing biomolecules. Focus on DNA bases complexes (quadruplex G4-DNA), ground state single particle electronic properties. Is DNA a viable electrical material? Experiments show DNA charge mobility is poor. For molecules attached to substrates. Use stiffer molecules or softer surfaces. Can also improve intrinsic conductivity through metal insertion. Metal-DNA interaction: different contexts - more interested in metals inside the helix. Guanine based DNA or metalized DNA. Use Density functional theory-GGA, ultrasoft psuedopotential, plane waves, etc. quadruple helices (G4-wires). X-ray structure is available: it is characterized by square and translation symmetries, K^+ are interplanar. The backbone is neglected in the simulation. Double ring of H-bonds, electronegative inner core, monovalent cations make G-wires thermodynamically stable, no in plane delocalized orbital bonding. K(I) G4 Electronic Structure: flat bands, no dispersion along wires axis, manifolds, effective semiconductor. Channels for charge motion through the bases, poor potassium-guanine coupling. Redox-active metals in G4-wires - Cu(I)G4: Cu integrates into plane. Cu(II) is more interesting, spin down state can accept electrons. Have played with different ion incorporation, but no successful experimental results. Metal incorporated wires - substitution of GC with hydroxypyridone. Spin density is a linear combination of spin up and spin down, strong coupling between metal and base. Electronic structure, 5 band manifolds originate peaks in the DOS, filled β-like HOMO, partially filled sigma orbitals, ferromagnetic alignment.

Francesco Gervasio (ETH Zurich) Charge transfer and oxidative damage in DNA fibers.
e-mail: fgervasi@phys.chem.ethz.ch

Guanine radical cation is the initial product of DNA oxidation by a wide variety of reagents: pulse radiolysis, photo-oxidants, transition metal complexes, UV. G^+ can be formed directly or via charge transfer. Charge transfer: the nature of DNA is a stack of aromatic molecules, possible conduction, highly insulating or superconducting depending on length. DNA is soft segmented, not a coherent long-range transfer. Charge hopping via tunneling (few bases) but is more likely polaron-like because of the title of the bases due to local distortion, or possibility of fluctuation of counter-ions, or deprotonation of G. Model system is the G:C decamer in the Z conformation. X-ray structure available, self-complementary. Crystal structure has few water molecules but is closely packed, no conformational changes, but possible loss of biological significance. Uses linear scaling methods to simplify calculations to laptop. From 300K to 0K polarons appear due to tilt of guanines. Third possibility is the G deprotonation: spin trapping. Simulations suggest that tilt of bases is not likely. H/D isotope effects. If the deprotonation was a parasite event an increase in the charge transfer is expected upon deuteration. In deuterated DNA the hole transfer is three times less efficient! At least the G:C couples the charge transfer is coupled to proton transfer. Reactions: Protonation State of the G:C couple, $\beta E \mu$ 3.6 kcal/mole, fate of G^+ in DNA. In duplex DNA the initial product is 8-oxoG. First step in oxidation reaction using ab-initio molecular dynamics. Rate limited by water autoprotolysis, catalyzed by phosphate backbone. Stabilization of the intermediate is achieved by proton transfer from the cytosine.

Bernd Giese (Basel) - Chemistry at a distance: Electron transport through DNA
e-mail: bernd.giese@unibas.ch
Ribonucleotide reductase, proteins have reaction taking place far away because of electron transfer. We are interested in the radical ion formation to radical ion transition through electron transfer in the DNA. Light generates a radical and we ask if charges stay or migrate. Analyze products where the charge is more stable. Examine gel electrophoresis of ^{32}P labeled oligos.
$\ln(K) \beta$ -$\beta\beta r$
where K is the rate constant, r the length of the electron transfer path, such that the decay constant is on the order of 0.7Å. The transfer stops because it is easier to go through water. Thus a change of mechanism is required so that distance influence no longer occurs and is likely that electron hopping is happening by some other mechanism. Note that injection point with cations does not have significant electron hopping dependency on length. For anion, how can we detect charge transfer? H trapping through thymine, verified by ESR. $K_{ET} = N^2$ which is a description of diffusion and is thus electron hopping. Tunneling is governed by $\ln(K)$.

Andrei Rakitin (York University, Toronto) DNA Electrical Properties: Natural Variety and Experimental Limits.
e-mail: rakitin@yorku.ca
C&E News, DNA: insulator or wire? (1997), DNA charge migration: no longer an issue (2001). The central issue is the difference is between charge transfer and charge

transferred. AT is an electron barrier for tunneling or thermal hopping. Charge transfer rate decreases exponentially, so that further distance does not change much. As orbital overlap increases then electron hopping dominates. LUMO/HOMO calculations suggests that DNA should be a wide band insulator. What is the density of states (DOS) at the Fermi level? Conductivity is determined by a narrow region near the contact where the voltage drops is on the order of $T\beta/e$. Tunneling, one contact (STM) actual band gap. Two contacts symmetric I-V E_f(metal) valence (DNA) E_f(metals). Gap in DOS indicates quasiparticle formation. How does DNA sequence effect electrical properties? Individual DNA is clearly wide band gap semiconductor. But DNA bundles have sequence disorder. Field effect: surface states – Fermi level pinning by interface states in a compound semiconductor is given as an example. Ferroelectric behavior, I-V behavior changes altogether. Conclusion: Question of DNA charge transfer is still open, new carefully controlled experiments required.

Alik Kasumov (RIKEN) Conductivity and induced Superconductivity in DNA

e-mail: kasumov@postman.riken.go.jp
Limits for Molecular Electronics: $Q = 50W/cm^2$ maximum heat removal, $N = 10^{12}/cm^2$ density of devices.
Graph of $Log[P(W)]$ vs. $Log[t_d(s)]$.
The Heisenberg principle restriction $E_{min} = h/t_d$, and
Von Neuman restriciton: $E_{min} = ln(2)kT$, T=300K.
Must resort to superconducting molecular computer to be able to handle energy dissipation, etc. Schematic picture of DNA is compressed at surfaces, even seen in TEM shadow images. Image simultaneous in topography and "spreading resistance mode". Without pentylamine treatment it is an insulator. With pentylamine it is a conductor. Used combing of β-DNA molecules to orient across IC slits. Low-temperature conductivity of DNA in differential voltage imaging. DNA molecules between metallic electrodes: if height of electrode is >3nm, cannot establish conductivity along the wires. Surface preparation (height of molecules) DNA molecules prepared in this way have resistances of μ 100 kβ/molecule. Proximity induced superconductivity in DNA molecules: contacts RE/C $T_c\mu$1K. DNA behaves like a doped semiconductor.

Claude Nogues (Weizmann Institute) Electrical properties of DNA characterized by conducting-atomic force microscopy

e-mail: claude@wisemail.weizmann.ac.il
DNA as a molecular wire: 1962 the issue of DNA conductivity was raised by Eley and Spivey, several other similar results though there is great disparity amongst the results. Important considerations: length of the DNA strand, probability of defects. Sequence – ionization potential G>A>C>T. Established contact between the DNA and the electrodes, to provide charge injection. Methodology, monolayer of thiolated ssDNA on a gold surface, monolayer of complementary ssDNA on gold nanoparticles (GNP), hybridization on surface, covalent bonding at both ends, only specific interactions.

We used a complex sequence of 26 bases with no chance for self-recognition. Samples prep: specificity of the adsorption via the thiol. Density: homogenous coverage, space for hybridization. Rinsing was undertaken to eliminate salt from ssDNA monolayer. GNP-dsDNA-Au bridge, and obtained good coverage on the surface, GNP marker for ds-DNA. Characterization: radio-labeled phosphate $3x10^{13}$ probes/cm^2, 96% bound to surface, when dsDNA is absorbed the height changes from 4 to 6 nm, GNP makes height changes from 10-14nm, believes only a few complementary strands bound. GNP complementary strands provides about 450 GNPs/μm^2, w/o complementary strand, about 5 GNPs/μm^2. Conductive AFM measurements with topography images: Tapping mode, bias 2V. Most of the GNPs are associated with an electrical signal. Control experiment, less GNP on the surface, no current detected. Formed mixed monolayer, 50% thiolated ssDNA, 50% mercaptoethanol. Current can go through by tunneling. GNP-covered with non-complementary ssDNA does not show a current, no ionic current. Scanned in tapping mode/I-V in contact mode and found conductive AFM I-V curves. Topography before and after AFM curves the GNP is shown to be pushed to the surface. Need to control the force to keep from squashing the particles into the surface. Upon GNP pressing, the blue line is approach, red is retract (force curve diagram). Chose location on force curve to minimize the pushing.

Danny Porath (Hebrew University) SPM and Charge Transport Measurement of Various DNA and DNA-based Molecules. Towards DNA-based Nanoelectronics?
e-mail: porath@chem.ch.huji.ac.il
Outline: Charge transport through standing DNA, G4-DNA, the puzzle of contrast inversion in DNA STM imaging. Three important features for DNA based nanoelectronics: recognition, structuring, and conductivity. Charge can be transported along short and single DNA polymers, in bundles and networks. Charge transport is blocked for long single DNA molecules that are attached to surfaces along their length. AFM Topography/Current Maps alternate complementary, non-complementary, complementary DNA strands with current flowing only through complementary strands. Current-voltage measurements: done during force distance (F-Z) curve. Green, jump into contact (JIC), pushing the GNP down, when tip is deformed we perform I-V, then retract (red) and adhesion layer holds tip onto surface (force curve is shown). In a controlled way we keep from pushing the GNP into the gold. Get non-linear I-V with over 200nA at 2V. Is this result real? I.e. is the conduction through the DNA itself? On cluster we believe yes, on ssDNA no current, back on gold, higher current. Then repeat by pressing with less force. Stretching the DNA, compare F-Z with I-Z. Done over a few nm, corresponding to a few pN. Current peaks after the tip is pulled back a bit, but drops when the DNA stretches. Cannot get results from ss DNA monolayer. Took on ssDNA simultaneous deflection and current versus bias voltage as a function of distance towards the surface. Little conductivity indicated in the I-V until the second jump into contact when you have Ohmic behavior. Experimental evidence appears strong that short strands do conduct. G4-DNA, it is a stiffer structure. Long poly G strand pH=13, slow pH reduction to pH 7 to get long monomolecular wire. Boiling at 100°C and treatment with DNase have

no observable effect on the molecules. Height is about double that for double stranded DNA.

Julio Gomez (Universidad Autonoma de Madrid) AFM based techniques to measure conductivity in the DNA based Nanowires.

e-mail: juliog@pop.uam.es

Nanowires overview: Break-junctions, carbon nanotubes, organic molecules, and quantum wires. SWNT can be described by two numbers, is therefore a crystal, and thus perfection is responsible for their unusual properties. Should the DNA be a conductor (poly(G)-poly(C)) get good bands, but one base out of sequence ruins symmetry. Mask with tungsten fiber covering DNA, metallized SFM tip. Electrostatic force in SFM: a non-intrusive method (conductive tip, cantilever, battery, and molecule). Molecule has both resistance and capacitance. Carbon nanotubes (CNTs) just about shine at edge because both gold and nanotube are conductive. Bias voltage about 2V. Nanotubes and DNA co-adsorbed nanotubes and DNA, CNT shine, DNA does not. DNA electrical properties without any electrical contact. Conducting molecule will not have electric field penetrating the tube (if over an insulating surface.) But it will penetrate the DNA. Comparison of force indicates CNT. High density poly(G)-poly(C) sample, no conductance, Increase humidity, saw that it improved. CNT glowed, but not DNA. Tried to do EFM on CNT and DNA, still CNT shines, DNA does not.

Hao Yan (Duke University) – New Structure for DNA-Based Nanofabrication

e-mail: hy1@cs.duke.edu

Background on why DNA: Predictable intermolecular interactions, convenient automated chemistry, convenient modifying enzymes, locally stiff polymer, and high function group density. Combine banded DNA with sticky ends to make objects and lattices. DNA templated self-assembly protein arrays, a new motif: cross structure. DNA nanoribbons: uncorrugate assembly – nanoribbons some times "bust open" and create 19nm arrays – cool molecules for imaging in the AFM. Scaffolding for 2D protein arrays through biotinilation of the DNA tile, the streptavidin will find to the biotin sites. Streptavidin is precisely located above the tile corners. Designed a two tile system such that only the "A" tile contains biotin, can not specially control the proteins with precise location. Switchable DNA lattices actuated by conformational changes: e.g. B-Z DNA switching devices, driven by ionic exchange or field strength. Motor motion is localized. Key challenges: incorporated DNA devices into self-assembled DNA lattices. Achieve switchable DNA lattice patterns. Potential applications, actuated assembly of nanoelectronics, control chemical synthesis templated on DNA tapestry, reversible pattern transition of protein folding. Nanoactuator has hairpin and linear state, imaged tapping in fluid with 10mM magnesium. Review fuel strand displacement used to drive DNA tweezers motor. Fuel strand removes "self strand", allows DNA to relax and combine with new complementary strand. Monitored with gel electrophoresis and fluorescence. Observed under AFM: autonomous unidirectional DNA walker – how to get it travel along a programmed route? DNA track is to follow anchorages with ligase, restriction enzyme and ATP. A hybridizes with B, knick is sealed, enzyme asymmetrically cleaves at A: A*B -> A +B*, can only go in one direction. Anchorages have hinges to allow them to bend. Detected using gel electrophoresis. No reverse step was allowed.

Marcus Helfrich (Penn State) – DNA nanoparticles: Towards Deterministic Bottom-up Assembly of Metal nanoparticles

e-mail: mrh14@psu.edu

Target architecture: nanowire raft. Synthesis of striped nanowires – templated electrodeposition generated in porous aluminum oxide. Nanowire electrical properties can be controlled by composition to make them conducting, semiconducting, and even photonconducting. Nanowire assembly at the oil/H_2O interface. 2D rafts form on the mm-scale, raft packing is determined by interparticle interactions. Switch to DNA as an assembly tool because of sequence-based selectivity and reversibility. DNA nanowire conjugates: Fluorescence-based biosensing on striped wires. DNA complementary strands excite fluorescence. Propose examining aqueous-aqueous interfaces – two immiscible polymer solutions, tunable properties/biocompatible, wires collect at the interface, use optical microscope "dipping lenses" for visualization. Rest of talk will focus on 320nm diameter gold wires. Complementary linkers make rafts but without linking strand, no wires, and they are finicky to produce. Nanowire raft formation at the APTS interface. How is wire mobility/hybridization influenced by colloid? Modeling wire behavior with 12nm colloid. Monitor hybridization by measuring color (plasmon) shift. Hybridization thermodynamics: polymers act as volume excluder; increase duplex stability. Chemical effect – PEG and Dextran impact T_m differently. Production is a balancing act between crowding and electrostatics. Colloid flocculates at high (15%) PEG concentrations in absence of a linker strand. Same effect can be seen in samples prepared in top phase of PEG 8 KDa Dextran and 100-200 KDa ATPS. Colloid flocculates above 6 wt% PEG in 8 KDA ATPS phase. Reduced salt concentration increases electrostatic repulsion. Can prevent non-specific flocculation through increase in electrostatic repulsion. The viscosity of ATPS: you can influence wire mobility by adjusting the viscosity of the polymer solutions. Dextran molecular weight has largest impact on viscosity. Interfacial tensions can be used to determine whether the wires will collect at the interface. Wire mobility at the interface.

Kalim Mir (Oxford University) – Spatially addressable self-assembly, combing and nanoparticle binding of single DNA polymers

e-mail: kalim@well.ox.ac.uk

Background – coming from the human genetics/genomics point of view. 4 billions bases, 30,000 genes, and several million sites of common variation. Need to link sequence/genes with traits/disease. Ordered array, spatially addressed (shows schematic of single chromosome). Diagnostic device "gene-chip" to sort complex mixtures of molecules. Extracted genome is disorganized, reorganized by hybridization on the chip. Microarray – synthesize probes and spot onto slide is one way, other approach is in-situ synthesis on a chip (photolithography). Digital microlithography aims: to develop massively parallel methods for genomic analysis in which single molecules can be viewed individually. Look at long range feature such as the genome and analyze short range feature of the genome. We want to sequence the human genome for $1000. A platform for single molecule analysis – sort, resolve, display, examine. Wants to be able to look into microarray spot and look at individual dye molecule markers. Can tell if single molecule by photobleaching characteristics. Use digital data extraction to count individual molecules and amplification of signal from single molecules using nanoparticle labeling. Self-assembly and combing of genomic DNA on microarrays, fragments sizes of up to 200kb, capture on microarray, comb, spot gives whole genome coverage that can be labeled with sequence specific probes. Haplotype: diploid genome, one copy from each parent, haplotype is the linear association of alleles on single parental chromosome. SNP typing does not resolve which set of alleles from which parent. Enables better correlation of genomic regions with disease. Haplotyping by spatial addressing: oligonucleotide nano-array, novel spatial arrangement of array elements identifies each allele the gene is from. Another way is to measure electrical continuity from each pad. Bridging DNA between probe coated electrodes: sequencing-by-synthesis on arrays, monitoring template-directed sequencing.

Tomoji Kawai (Osaka) - DNA based electric and magnetic devices

e-mail: kawai@sanken.asaka-u.ac.jp
Seeks to make bottom-up biomolecular devices. What kind of electronic properties and structures does DNA have? DNA is a very good scaffold for construction, 2nm width, 1D chain. Expect conductivity because of base stacking, but no consensus on its conductive properties. Placed Poly(dG)Poly(dC) on SiO_2/Si. XPS spectra, 5 eV difference between HOMO and LUMO, appears to acts as a wide gap semiconductor. To control electrical conductivity, must dope with hole or electron injection. Chemical, electrical or photo-doping. Scheme of the controlled conjugation of Au nanoparticles. Complementary strands with cobalt particles make nanoscale magnetic memory. Recently made 5-nm gold nanoparticle arrays on surface, using stepped surface of sapphire (0001) substrate and gold particles align with DNA and steps.

James Vesenka (University of New England) - Auto-orientation of "G-wire" DNA

e-mail: jvesenka@une.edu
Objective: to understand the apparent auto-orientation of "G-wire" DNA on the surface of mica. G-wires, named by Tom Marsh (Marsh, Vesenka, & Henderson, *N.A.R.* 23, 696-700 (1995)), consist of four G-quartet repeats with thymine linkers, can reach micrometers in length. They appear to be very uniform on the surface of mica compared to double stranded DNA and don't collapse on the surface of mica (T. Muir, e. al., *J. Vac. Sci. Technol. A.* 16, 1172-1177 (1998)). The curious observation is that the G-wires appear to orient at 60° intervals on the surface of mica. Comparison with the underlying substrate suggests that the G-wires align with next

nearest neighbor potassium vacancies through a divalent cation tether (typically magnesium) to the phosphate backbone of the G-wires. Sometimes the G-wires orient in three directions, sometimes only one. We speculate this is a result of the quality of the mica and the way the potassium ions vacate the surface of mica upon cleaving. The potassium may vacate in straight rows, staggered rows, or as clusters, enabling certain preferred orientations of the G-wires on the mica surface (i.e. in three directions or only one.) A control study using APTES to bind the DNA to mica indicates random orientation of the DNA over the surface, expected since the DNA no longer will have a preferred orientation because of the randomized surface tethering. Possible applications in providing nanoelectronic templates or possibly even nanowires, though conductivity, as indicated by low current scanning tunneling microscopy is only hinted at.

Keith Williams (Delft) Assembly of Nanotube-based Electronic Devices by Biomolecular Recognition
e-mail: K.A. Williams@tnw.tudelft.nl
Nanotube, DNA linker, thiol linker to gold contacts: most important feature is "reconfigurability". Getting nanotubes to stick to substrate is easy, to come off has not been solved. Covalent chemistry approach with the following motivation: miniaturization, hands-free bottom up assembly with a high degree of perfection. Focus on bio-aspects of assembly: biology provides unique tools for assembly, ordering, and replication. The interface between biology and physics is interesting, and needs to be explored in greater depth. Introduction to PNA properties: "pseudo-peptide" backbone has no charge unlike the phosphate groups. PNA can have solubility problems and higher T_m than DNA. Single stranded DNA likes to wrap around nanotubes. Amide linkage of PNA to nanotubes: undertake Watson-Crick hybridization of DNA to PNA adducts on nanotubes. Present AFM evidence for the tubes, noteworthy that DNA, though larger than tubes ends up collapsing on mica. Current, optimized amide linkage of PNA to Nano tubes improves the yield. The "nanosperm": an attachment at one end of the nanotube with the DNA. Possible reasons for these effects include nonspecific binding, side reactions with exocyclic amines, water-solubility spacer needed. SWNT attachment to Protected PNA-Resin: adducts can be counted by UV to test for loading of FMOC, now in process, fluorescence labeling. Now, attaching onto gold and hybridize onto the DNA. Also looking at intermediate linkers such as gold nanoparticles, evaluated with physical absorption.

Jürgen Schlütter (LOT Darmstadt) Dip Pen Nanolithography: A new tool for the fabrication of nanostructures.
e-mail: schluetter@lot-oriel.de
AFM tip coated with molecules, water meniscus develops between and substrate such that a nanopattern can be constructed (Hans Jurgen Butt '96, Chad Mirkin '99). Scalability, can work with multiple pens. Thiols and gold, linewidth of about 90nm, variation less than 10%. Photolithography is the workhorse of IC industry but has a

practical limit at about 100nm. E-beam can create structure down to 5-10nm, nano-imprint also 10nm. Seek a bottom-up approach to the 5-10nm limit. Direct write techniques enable deposition of soft and hard nanostructures, with a spatial resolution of as small as 5nm, but typically 30-40nm. SPM does alignment, writing, and inspection. General molecule and substrate considerations: highly scalable with parallel pens systems (multiple AFM tips), passive arrays. Actively controlled tip arrays are more interesting with active ink delivery. Experimental factors affecting DPN: temperature increases with motion, humidity (stable between 30-60%), time (determined by tip speed, dot size increases with contact time). DPN Inks: soft materials – small functional molecules (dyes), SAMS, conducting polymers, and biopolymers (DNA, proteins). Hard materials include metal inks and solid precursors. Etch barriers for solid-state nanostructures (write regions of octadecathiol on Au thin film. Selectively etch Au using wet chemical etchant, remove Ti and SiO_2. Etch resists for nanofabrication, 60nm wide, 12nm nanogap. DNA-functionalized nanoparticle device to fit in the gap and characterize the electrical nature of the structure. Template driven assembly via DPN, to make combinatorial DPN templates – millions of experiments on one area of a substrate. Carbon Nanotube and Nanowire Devices via DPN. Microfabricated electrodes, glue on molecular layer, assemble nanotube. Complimentary with microlithograpyhy: design pattern, print and inspect, cannot absorb nanoparticles, nanotubes. DNA as ink for nanolithography. Direct-write patterns on SiO_x, direct wrote ssDNA, complementary strands with different fluorophores gave different colored signals. Optical techniques are the draw back. Replace with 25 and 13nm gold. Took about 20 min to 400nm x 16 µm array with 8 pens with AFM stage.

PHOTOS (1)

Come together
at the Optical Museum
(Thursday, May 13)

Talks (Friday, May 14)

N. Seeman (New York)

I. Willner (Jerusalem)

A. Woolley (Young Univ.)

A. Filoramo (Saclay)

W. Fritzsche (Jena)

M. Washizu (Tokyo)

M. Heller (San Diego)

A. Estabrook (UCSB)

R. Di Felice (Modena)

R. Gervasio (Lugano)

B. Giese (Basel)

A. Rakitin (Toronto)

PHOTOS (2)

Talks (Friday, May 14; cont.)

A. Kasumov (RIKEN)   C. Nogues (Weizmann)   D. Porath (Jerusalem)   J. Gomez (Madrid)

Orchid excursion (Friday, May 14)

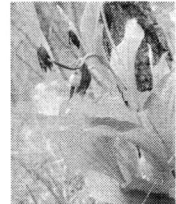 

Talks (Saturday, May 15)

M. Helfrich (Penn State)   K. Mir (Oxford)   T. Kawai (Osaka)   J. Vesenka (Uni New Engl.)

H. Yan (Duke Univ.)   K. Williams (Delft)   J. Schlütter (Darmstadt)

112

LIST OF PARTICIPANTS

Arinaga, Kenij	Munich	arinaga@wsi.tum.de
Bier, Frank	Potsdam	frank.bier@ibmt.fhg.de
Bretschneider, Jan	Aachen	jan.bretschneider@ac.rwth-aachen.de
Bulla, Ralf	Augsburg	Ralf.Bulla@Physik.Uni-Augsburg.De
Ceyhan, Bülent	Dortmund	ceyhan@chemie.uni-dortmund.de
Csaki, Andrea	Jena	csaki@ipht-jena.de
Dettenwanger, Franz	Hannover	dettenwanger@volkswagenstiftung.de
Di Felice, Rosa	Modena	rosa@unimore.it
Estabrook, August	Santa Barbara	aestabrook@chem.ucsb.edu
Festag, Grit	Jena	grit.festag@ipht-jena.de
Filorama, Arianna	Saclay France	filoramo@drecam.saclay.cea.fr
Fischler, Monika	Aachen	monika.fischler@ac.rwth-aachen.de
Fritzsche, Wolfgang	Jena	fritzsche@ipht-jena.de
Gervasio, Francesco	Zuerich	fgervasi@phys.chem.ethz.ch
Giese, Bernd	Basel	bernd.giese@unibas.ch
Gloddek, Kirsten	Aachen	KIRSTEN.GLODDEK@ac.rwth-aachen.de
Gomez, Julio	Madrid	juliog@pop.uam.es
Helfrich, Marcus	Penn State	mrh14@psu.edu
Heller, Michael	San Diego	mjheller@ucsd.edu
Hesse, Eva	Aachen	eva.hesse@ac.rwth-aachen.de
Holste, Claudia	Jena	claudia.holste@ipht-jena.de
Inganäs, Olle	Linköping	ois@ifm.liu.se
Kasumov, Alik	RIKEN Japan	kasumov@postman.riken.go.jp
Kawai, Tomoji	Osaka	kawai@sanken.osaka-u.ac.jp
Klinov, Dmitr	Moscow	Klinov@mail.ibch.ru
Kretschmer, Robert	Jena	robert.kretschmer@ipht-jena.de
Mir, Kalim	Oxford	kalim@well.ox.ac.uk
Nogues, Claude	Weizmann Inst.	Claude@wisemail.weizmann.ac.il
Noyong, Michael	Aachen	michael.noyong@ac.rwth-aachen.de
Porath, Danny	Jerusalem	porath@chem.ch.huji.ac.il
Rakitin, Andrei	Toronto	rakitin@yorku.ca
Schlütter, Jörg	Darmstadt	schluetter@lot-oriel.de
Seeman, Nadrian	New York	ncs1@feynman.acf.nyu.edu
Steinbrück, Andrea	Jena	andrea.steinbrueck@ipht-jena.de
Tanaka, Shin-ichi	Osaka	t-shin32@sanken.osaka-u.ac.jp
Vesenka, James	Univ. of New England	jvesenka@une.edu
Wagner, Jörg	Ilmenau	Joerg.Wagner@Tu-Ilmenau.De
Washizu, Masao	Tokyo	washizu@washizu.t.u-tokyo.ac.jp
Williams, Keith	Delft	K.A.Williams@tnw.tudelft.nl
Willner, Itamar	Jerusalem	willnea@vms.huji.ac.il
Wolff, Andreas	Jena	andreas.wolff@ipht-jena.de
Woolley, Adam	Brigham Young Univ.	awoolley@chem.byu.edu
Yan, Hao	Duke University	hy1@cs.duke.edu

AUTHOR INDEX